OLD BONES AND SERPENT STONES

Old Bones

— AND —

Serpent Stones

A Guide to Interpreted Fossil Localities
in Canada and the United States
Volume 2: Western Sites

by
T. Skwara

with line drawings
by
I. Fay

The McDonald & Woodward Publishing Company
Blacksburg, Virginia
1992

The McDonald and Woodward Publishing Company
P. O. Box 10308, Blacksburg, Virginia 24062–0308

Guides to the American Landscape

**Old Bones and Serpent Stones: A Guide to Interpreted Fossil
Localities in Canada and the United States
Volume 2: Western Sites**

© 1992 by The McDonald and Woodward Publishing Company.

All Rights Reserved
Composition by Marathon Typesetting, Roanoke, Virginia
Printed in the United States of America by McNaughton and Gunn, Inc.,
Saline, Michigan

99 98 97 96 95 94 93 92   10 9 8 7 6 5 4 3 2 1

First Printing, April 1992

Library of Congress Cataloging-in-Publication Data

Old bones and serpent stones : a guide to interpreted fossil localities in
    Canada and the United States.
        p.   cm. -- (Guides to the American landscape)
        Includes bibliographical references and indexes.
    Contents: v. 1. Eastern sites / by Jerry N. McDonald -- v. 2. Western
    sites / by T. Skwara.
    ISBN 0-939923-08-4 (v. 1) : $14.95. -- ISBN 0- 939923-09-2 (v. 2) :
$14.95.
    1. Paleontology--Canada--Guide-books. 2. Paleontology--United
States--Guide-books. 3. Fossils--Canada. 4. Fossils--United States. I.
McDonald, Jerry N. 1944-   . II. Skwara, Theresa. 1949-   . III.
Series: McDonald & Woodward guide to the American landscape.
QE705.C2045   1991   91-19951
                                                               CIP

Acknowledgments

Many people associated with museums, parks, and fossil sites in western Canada and United States generously provided information when I was compiling the data for *Old Bones and Serpent Stones;* others reviewed appropriate portions of the penultimate draft of the manuscript. I am most grateful for their assistance. I am particularly grateful to the library staff at the American Geographical Society Collection, University of Wisconsin at Milwaukee Library for facilitating access to topographic maps via inter-library loans. Dr. Michael R. Voorhies, Curator of Vertebrate Paleontology, University of Nebraska State Museum, provided up-to-date information on the new Ashfall Fossil Beds State Historical Park in Nebraska.

I have relied heavily on the research and writings of many paleontologists and geologists. Without that collected body of knowledge, I could not have written *Old Bones and Serpent Stones.* I acknowledge the research of the scientists and writers whose works are cited in Section III. The friends and colleagues who provided invaluable criticism and encouragement prefer to remain anonymous, and I will respect their wishes. However, I do want to acknowledge my parents for their unstinting support of this undertaking.

Jerry McDonald, collaborator in *Old Bones and Serpent Stones (Volume 1: Eastern Sites)* has become a friend.

Contents

Introduction

There is a special thrill that comes from studying fossils, from the view of ancient worlds that they afford us, from the understanding we gain of past, present, and even future time. This thrill can only be experienced by those willing to venture into the vastness of deep time[1] and to face the implications of deep time for human self-awareness.

In recent years, there has been a tremendous resurgence of interest in fossils among people of all ages; for young people in particular, dinosaurs have been the focus. Among scientists, too, interest in paleontology has been revived, and the role of paleontology is being reevaluated. No longer viewed as mere stamp collectors of the scientific world, paleontologists are seen providing unique insights into the complex interaction of evolving biosphere and lithosphere, atmosphere and hydrosphere.

Public interest in paleontology is fueling a demand for greater access to and information about fossils. But the voices of most paleontologists emanate from within the hallowed halls of institutions of higher learning and are often shrouded in arcane terminology. The present book is an attempt to bridge the gap between the two groups of fossil lovers, each curious about the history of life on earth. It is an invitation to explore deep time, to discover incredibly ancient worlds inhabited by fascinating organisms long since extinct.

---

[1]This appropriately emotive phrase from John McPhee, 1980, *Basin and Range*, has very quickly become incorporated into the vernacular of geology and paleontology.

1

Scope and Purpose

Old Bones and Serpent Stones is written with two objectives in mind. It is primarily a guide to those fossil sites in western Canada and United States that are accessible to the public and, as such, is intended for the use of people who are, or will be, visiting those localities. The book is also intended for a general readership as a source of information about paleontology. In either case, it is meant to provide access to the often rarified atmosphere of science.

The general organization of the book reflects the duality of purpose. In Section I, various aspects of paleontology are examined; a discussion of concepts is followed by an historical account of life on earth and of the geology of North America, with emphasis on western North America. The topics that are presented reflect the interdisciplinary nature of the science of paleontology, and the sites that are listed as examples are ones that are described in Section II. In Section II, the fossil sites in western Canada and United States that are open to the public are documented. The sites are described one at a time, with both paleontological interpretation and practical information provided, and are arranged according to their geographical distribution.

Sources of additional information are compiled in a separate section. In particular, Section III includes a selected bibliography on topics in paleontology. The citations are intended to complement as well as supplement the concepts and interpretations presented in sections I and II. Some of the references cited are broad in scope; others are more specialized and can be used as stepping stones into the scientific literature. The reader is also referred to museums and field courses for a broader range of experiences.

The sites that are documented in *Old Bones and Serpent Stones* were selected according to three criteria:

1) Interpretation—Is there some source of information about the fossils and the fossil site readily available to the public, either at the site or in the vicinity?

2) Protection—Does the site have legal protection? Is the site protected from vandalism in any real, physical way that reinforces the legal protection?

3) Accessibility—Is the site open to the public? Is there reasonable public access?

Each site documented in Section II offers interpretation, each is protected, and each is accessible. The quality of the sites, the level of development, the degree and type of interpretation, and the access to fossils, however, are highly variable. Because the focus of the book is interpretive, sites where collecting is the focus and is encouraged, promoted, or facilitated—even where the collecting is legal—were deliberately excluded.

Caveat on Fossil Collecting

Old Bones and Serpent Stones is a book about fossils. It is not, however, a book about collecting fossils. All of the sites that are documented are protected by law, and fossil collecting at any of them is strictly prohibited.

All but three sites are located on lands owned or managed by public agencies, and general public access is usually unrestricted. Visitors who go beyond the areas developed for fossil exhibit, especially hikers in back-country areas, are likely to see fossils weathering out of outcrop. Such discoveries should be reported to park rangers or agency staff who will undertake the responsibility for correctly collecting the fossil. In some cases, park rangers will encourage visitors to participate in the collecting process. On private land, the permission of the land owner is required both for access and for collecting. Nonetheless, in some jurisdictions, fossils that are found on private land or uncontrolled public land may belong to public institutions (e.g., state, provincial, or university museums).

Fossils are a natural resource, and they constitute a legacy for all of us. Scientific and educational value, however, accrues only to a fossil that has been properly documented and collected; otherwise, it is a mere curiosity. The laws controlling fossil collecting are intended to protect that value.

Section I

The Many Faces of Paleontology

Fossils in isolation are antiquarian objects; in the absence of context and concepts they are mute. But with conceptual models and technical tools at our disposal, a rich and luxuriant tapestry—the history of life on earth—emerges. Fossils, time, and change are the foundations of that history.

The Background

About Fossils . . .

A fossil is a bit of bone glistening in the sunlight on a badlands slope. A fossil is a shell frozen in time on a limestone ledge high on a mountain slope, so perfectly preserved you have to pinch yourself to remember that this is not a beach. Footprints of giants on ancient beaches, burrows dug by anonymous worms, coprolites (fossilized feces), amber amulets, fossil fuels: these, too, are fossils.

Fossils fascinate us. They compel us. They inspire us and teach us. And yet, do we know what they are? Do we know how to find them? And where?

Fossils are the remains or traces of once-living organisms that are preserved in the rocks and sediments of the earth. They occur in a spectacular range of sizes and shapes and are preserved in a number of ways, some more common than others, that reflect the local circumstances at the time the organism died. Fossils are always distinct in some way from the other constituents of the rocks or sediments of which they are parts.

Most fossils are found in sediments or sedimentary rocks, layered deposits composed of sand, silt, mud, and organic debris. Such particles can be deposited on land or in the ocean, and the layers of sediment that accumulate may eventually be altered to form sedimentary rocks. Marine environments are almost universally sites of deposition, terrestrial environments more often the sites of erosion. Furthermore, oceans have always covered large portions of the earth, frequently flooding the continents. It follows, therefore, that most sedimentary rocks are of marine origin, and most of the fossils that we know are found in marine deposits. Nonetheless, many terrestrial settings, such as stream channels and lakes, are sites of deposition; the sediments that accumulate house the fossils of land plants and animals.

Fossils are rarely preserved in igneous or metamorphic rocks. The extreme heat that forms such rocks usually destroys the evidence of any organic material that may have been present. There are exceptions, however; the historical city of Pompeii, which was destroyed and its inhabitants entombed by voluminous ashfalls when nearby Mt. Vesuvius erupted in A.D. 79, provides the analogy. Once in a while, ancient organisms are found preserved in volcanic rocks because unusual circumstances mitigated the extreme heat of eruptions. At Ginkgo Petrified Forest State Park in Washington [Site 6], lava flowed into lakes and was sufficiently cooled so that logs and stumps in the lakes were preserved in lava. In other places, ashfalls from volcanic eruptions settling in a lake were particularly effective in preserving fragile insects and plants in minute detail. This mechanism explains the beautiful preservation of delicate organisms at Florissant Fossil Beds National Monument in Colorado [Site 22].

Fossilization is fortuitous and rare, but it is not unusual. Chance and circumstance determine whether organic remains will be preserved; nonetheless, the nature of the organisms themselves can alter the odds of fossilization. The ultimate fate of most organisms is decomposition, soft parts decomposed by the action of other organisms, particularly bacteria and fungi. It is unlikely that any organic remains will remain intact. Yet, if an organism has hard parts, a shell or bones, the chances that it will be preserved are greater than those of a completely soft-bodied organism. Even so, hard parts are destroyed by chemical activity and mechanical abrasion. Thousands of shells, for example, are battered every day by waves breaking on a beach. Thus, the environment in which an organism lived is, in many cases, the determining factor in fossilization; rapid burial by sands and muds of an organism with hard parts is nature's surest way of producing a fossil.[2]

Unusual circumstances and special conditions play an important role in the process of fossilization. The fossil record is replete with examples of spectacular preservation in unique settings: three-dimensional anatomical details of soft-bodied organisms found in the Burgess Shale in British Columbia [Site 4], wasps and butterflies in volcanic ash at Florissant [Site 22], ice-age mammal bones preserved in tar at Rancho La Brea in California [Site 43]. Fossils preserved in such ways offer detailed views of and rare insight into life in ancient time.

Fossilization generally occurs in a limited number of ways; in most cases, only the hard parts of organisms are preserved (Figure 1). Only rarely do fossils retain exactly the same chemical composition that the living organisms had (Figure 1a). In other words, the original shell or

---

[2]The study of the post-mortem history of plant and animal remains forms a branch of paleontology called taphonomy. It seeks to understand the events that occurred after the death of an organism and the circumstances of the eventual burial and preservation of a fossil. It is important to point out that all fossils reflect both the biological community in which the organism lived and the taphonomic processes that modified the organic remains after death.

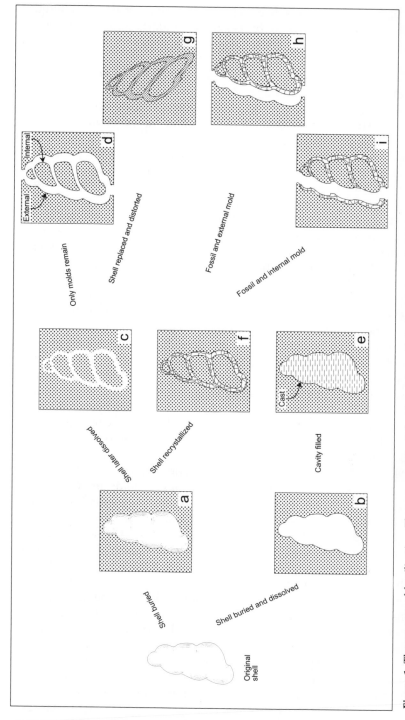

Figure 1. The process of fossilization. The cross-sectional drawings of a snail shell show how original hard parts of organisms are usually altered during the process of fossilization. Figures **a** to **i** are explained in the text.

13

bone, be it calcium carbonate or calcium phosphate, is usually altered in the fossil, the original hard parts having undergone some degree of chemical or mineralogical alteration over the period of time since they were first deposited.

Organic remains can be altered in a number of ways. Many hard parts are simply buried and dissolved (Figure 1b, 1c, 1d), in many cases the remaining void filled with different material (Figure 1e). The original crystalline structure of organic hard parts may be recrystallized (Figure 1f). Many shelly organisms during life construct solid shells by depositing calcium carbonate in a crystal form called aragonite; during recrystallization the aragonite is altered to a new form of calcium carbonate called calcite, and the microscopic detail of the fossil is destroyed. Replacement may take place; the original material of the hard parts is dissolved and replaced, molecule-by-molecule, by another mineral (Figure 1g). If the replacement mineral is silica or pyrite, the process will preserve even the finest microscopic details of the original. In cases such as these, the fossils retain the form of the original organism (Figure 1h, 1i), but in other cases, the fossil that remains is preserved as a mold or cast of the original.

Bone and wood are usually preserved in other ways. Because they are very porous organic hard parts, bone and wood are readily altered by permineralization, the petrification process whereby minerals such as silica dissolved in groundwater are deposited in the pores. In some cases, the silica completely encloses the organic material and protects it from decay. The deposition of minerals greatly increases the weight of the original bone or wood. Alternatively, vegetative and bony remains can be preserved by carbonization. As the sediments that enclose the organic remains are compacted, the volatile organic components are driven off, and a thin film of black carbon remains on the rock. Minute details of many leaves and fish are preserved in this way.

Most fossils represent, in one way or another, actual parts of organisms: leaves, bones, or shells; the originals or the molds and casts. Trace fossils, in contrast, are an entirely distinct category of fossils. They constitute evidence of the past activity of organisms; no portion of the organism itself is preserved. That evidence includes tracks, traces, burrows, borings, coprolites, and gastroliths (stomach stones). Dinosaur footprints, for example, impressive and well known trace fossils, are important because they can provide clues about animal behavior that cannot be determined by looking at the bones alone. Some of the first interpretations of dinosaur behavior, such as speed of movement and the presence of family groups or herds, were based on footprints found along the Paluxy River in what is now Dinosaur Valley State Park in Texas [Site 28]. Invertebrate organisms, too, leave trails behind as they rest on the sea floor or drag themselves along the soft sediment searching for and consuming food. Invertebrate traces, common in certain settings, are well preserved at the Indian Springs Trace Fossil Site in Colorado [Site 24].

14

Human beings have left their tracks and traces on the surface of the earth, tracks and traces that properly constitute fossils. The study of the evidence of human activity falls within the purview of archeology. Usually, the distinction between paleontology and archeology is clear: remains of extinct plants and animals fall within the scope of paleontology, human artifacts and evidence of human activity within the scope of archeology. Nonetheless, there is an overlap where evidence of human activity coexists with remains of extinct organisms. A case in point is a now famous discovery at Blackwater Draw near Clovis, New Mexico [Site 26]: unique spear points, fashioned by skilled knappers, found embedded in bones of mammoths and bison, species of each now extinct. Here was the archeological proof, from the closing days of the last Ice Age, of big-game hunting societies in North America, of a time when humans coexisted with now-extinct animals. Here was also fossil evidence that might explain, in whole or in part, the extinction of big-game mammals in North America at the end of the last Ice Age. Perhaps the highly skilled new predator, *Homo sapiens*, who immigrated into North America 15,000 to 11,000 years ago, hunted to excess the big-game animals they relied upon for food.

A fossil can be one of many things. Yet whatever form it takes, it provides a view of an organism now extinct living in a time long past.

. . . and the Study of Fossils

People through the ages have considered fossils special objects and have attributed magical powers to them. Fossils capture human imagination: to some people they are sacred totems; to others, powerful aphrodisiacs; to still others, evidence of the biblical flood—the anger of the Judean god. The appeal of fossils is universal. On the one hand, our aesthetic sensibilities are triggered because, surely, fossils can be beautiful, and the wonder is that any fossils should have been preserved at all. On the other, our mind is challenged, our curiosity awakened, for fossils are cloaked in mysteries that intrigue every fossil lover. Where do fossils come from? How do they get to be preserved? How old are they? What do they mean? What can they tell us about the past?

Fossils provide unique glimpses into the history of the earth; like a time machine, each one transports us back to a unique time in the distant past. They are excellent time clocks, indicators of particular segments of geological time. Fossils are the proof of the evolution of life on earth, of the existence of organisms now extinct. Many fossils are precise indicators of environmental conditions that existed in the past; others document ancient geography, recording the formation and breakup of continents through the ages. By putting together the pieces of the puzzle that each fossil affords, we can reconstruct the history of life.

15

Fossils can provide answers to the questions we ask only if the fossils have been meticulously collected and well recorded. The interpretive value of a fossil lies in its context; the fossil itself is far less informative when data about its geographical and geological setting are lost. It is not just the fossil, but where it was found and the nature of the sediments that enclosed it, that is important. Dinosaur bones from bonebeds at Dinosaur National Monument in Utah [Site 40], for example, are different in every conceivable way from dinosaur bones found in bonebeds in Dinosaur Provincial Park in Alberta [Site 2].

Fossil collecting involves two distinct steps. The first is documentation, recording the geographical setting of the fossil, the type of rock in which it occurs, the characteristic features of the rock that might reveal how it was formed, the relationships among the various types of rocks that are present in the area, and so on. Only then is the second step—removal—undertaken. Usually just enough of the fossil is exposed in the field to show what is present and how large it is. The rock surrounding and enclosing the fossil is cut out, and the entire block is removed; then it is taken to a laboratory where the fossils can be exhumed more carefully.

Fossil collectors are pragmatic. The techniques they use vary from one type of fossil to another and from one locality to another. Some fossils are more easily collected than others; shells, for example, are usually easier to collect than bones. Some bones are better preserved than others. Some fossils are so small that they are invisible in the field setting; instead, buckets of sand or blocks of rock are collected and taken to the laboratory, where the material is examined under a microscope in the hope that it contains tiny fossils. It is also true that some fossils have more interpretive value than others, rarity often enhancing scientific value.

Some fossils are economically important. The fossil fuels are the most significant of these, for on them the wealth of the modern industrial economy is based. The role of microfossils, the remains of microscopic organisms, in both the formation of and exploration for fossil fuels, cannot be overstated. Some fossils, in particular petrified wood and dinosaur bones, can act as traps for uranium mineralization. Other fossils, in the form of blocks hewn from fossiliferous limestone are used as building stone to beautify our cities and, if metamorphosed to marble, are sculpted into works of art that grace art galleries and museums.

Fossils, beautiful and mysterious, will continue to fascinate us. So we study them, driven by our need to know, to understand, to read the cryptic messages they have written in the rocks.

Concepts in Paleontology

paleontology (pā'lē-on-tol'ə-jē). n, [<Gr. *palaios,* ancient; +
Gr. *onta, ontos,* a being, a living thing; + Gr. *logos,* the study
of] the study of ancient living things.

Paleontology is the science—and the art—of studying ancient life
on earth and the changing forms of life through time. It is a unique
discipline, for the study of life through time is accomplished by complex
integration of the two great concepts of science: geological time and biological evolution. Both concepts have had an impact far beyond the realm
of science; each has irrevocably altered the perception humanity has of
itself and of its place in the natural world.

Geological Time

Human beings have always been mesmerized by time, and as
human cultures evolved so did conceptions of time and the methods by
which people marked its passage. Thus, in some cultures, time is cyclic;
in others, it stops periodically. In western culture, time is linear and
directional. To mark the passage of time, we note the motion of the moon
and sun across the sky and use digital, nuclear clocks as our standard
of measurement. Nonetheless, time remains "the most elusive and mysterious of the primitive dimensions of [human] experiences."[3] These experiences dictate that we measure time in terms of human generations.

Geology shattered the comfort of human-bound concepts of time,
anthropocentric concepts equating the age of the earth with the age of
the culture that does the measuring. The age of the universe is no longer
equivalent to the age of human culture, and time is no longer synony-
mous with age. Geology, by measuring the age of an immensely old
earth, has given us a concept of time that is independent of ourselves.
Time—geological time—is the most important contribution that geology
has made to human thought.

The passage of time can be documented by either absolute or relative
dating. Absolute documentation is precise and accurate according to a
numerical scale. For example, the eruption of Mount St. Helens, a geological event, occurred on May 18, 1980, the date affixed on the accepted
standard numerical scale. Relative dating, in contrast, marks the passage
of time by the order in which unique events occur. Thus, human memory
recalls that Mount St. Helens erupted before an earthquake damaged
Mexico City even if memory cannot recall the absolute dates of each
event.

---

[3]Daniel J. Boorstin, 1983, *The Discoverers,* p. xvi.

Geological time is measured by both absolute and relative dating methods, and the geological time scale as we now know it is the result. While it is true that we can now accurately date the age of the earth and many of the events in earth history, for most aspects of paleontology and geology it is not the absolute age but the vastness of time and the relative timing of events that is important. The geological time scale reflects the vagaries of usage: providing a scale in millions of years, it allows us to date events; but it also subdivides time into eons, eras, and periods, each characterized by a unique sequence of events.

Relative geological time was first conceptualized some 300 years ago, and several fundamental principles of geology were defined. Two were elucidated by Nicholas Steno, a Danish physician working in the de Medici court in Florence. In *The Prodromus*, published in 1668, he established that fossils—sports of nature, or form stones as they were then called—were the remains of once-living organisms. Specifically, he showed that a particular kind of form stone, called a tongue stone, collected in vast numbers in Malta was, in fact, a tooth of a once-living shark. Steno also established what geologists today call the Principle of Superposition of Strata. It means simply, that in any sequence of layered rocks, the rocks at the bottom of the sequence were deposited first and are, therefore, the oldest (Figure 2a). A trivial analogy illustrates the principle: the floor boards in a house are laid before are, therefore, older than the carpet that covers them. The importance of the principle of superposition for conceptualizing geological time cannot be overstated, for it allows us to infer time from layers of rocks.

William Smith, an English surveyor of mines and canals in the late eighteenth and early nineteenth centuries, recognized a critical relationship between layered rocks and the fossils they contained. He observed that a discrete group of fossils occurs in distinct layers of sedimentary rocks, and he recognized that these groups of fossils succeed each other in an orderly fashion. This is the Principle of Faunal Succession: each rock layer has unique fossils, the order of the fossils one on top of another is constant, and the sequence acts as a universal standard (Figure 2b). Fossiliferous rocks from any region could now be studied and placed in their correct relative position. From the Principle of Superposition of Strata, the correct relative age could also be determined. The implications are tremendous! The Principle of Faunal Succession means that rocks from different regions that have the same groups of fossils can be correlated. More than that, it means that rocks having the same fossils must be the same age (Figure 3).

Geological time, as we define it today, is subdivided on the bases of unique events in earth history. Three broad subdivisions are recognized: the Archean and the Proterozoic, together encompassing more than 85 percent of geological time and including rocks that are predominantly

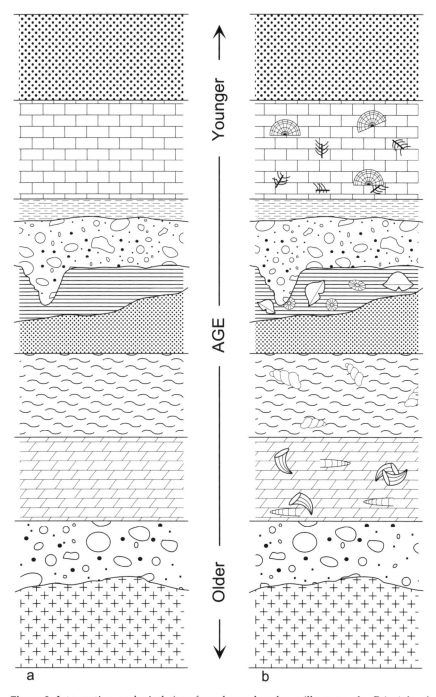

Figure 2. Interpreting geological time from layered rocks: **a** illustrates the Principle of Superposition of Strata; **b**, the Principle of Faunal Succession.

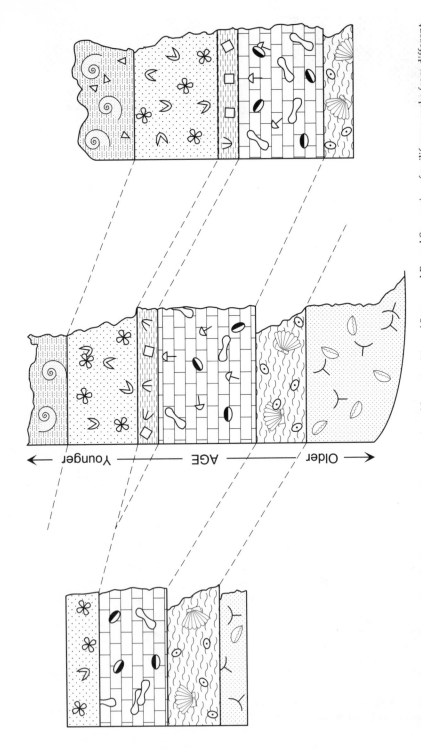

Figure 3. The correlation of rocks. On the basis of the principles of Superposition of Strata and Faunal Succession, fossiliferous rocks from different regions can be correlated. The dashed lines indicate fossiliferous rocks of equivalent age.

20

volcanic and metamorphic containing few fossils, are frequently combined and referred to informally as the Precambrian; the Phanerozoic is the youngest subdivision and is represented, for the most part, by fossil-bearing sedimentary rocks. Abundant evidence of life, in the form of fossils in the rocks, marks the beginning of Phanerozoic time, which is then subdivided into three eras on the bases of distinct groups of fossils and the order in which they occur in rocks. A relative geological time scale is thereby established (Figure 4).

The modern geological time scale reflects current knowledge of both relative and absolute dating methods. Relative dating provides the basic subdivisions of geological time; absolute dating, the ages in years of the various subdivisions. The age of the earth—the beginning of Archean time—is 4.6 billion years.

The oldest subdivision of Phanerozoic time by virtue of the position of the rocks at the bottom of the pile is the Paleozoic Era, the age of ancient life, characterized by fossils of ancient marine life. The Mesozoic Era is younger; the rocks overlie the rocks deposited during the Paleozoic Era, and the fossils represent middle life, the age of the dinosaurs. The youngest is the Cenozoic Era; the fossils define modern life, the age of the mammals. The boundaries between the eras are marked by mass extinctions, great disruptions in the nature of life. Each era, too, is subdivided into periods because discrete and unique assemblages of fossils within the rocks allow finer subdivision. The boundaries between the periods are marked by the extinction of characteristic organisms and the rise of new and distinct ones in the succeeding period. The names of many of the periods reflect the localities where the rocks and the unique fossils contained in them were first recognized. Devonian, for example, is named for Devon in England; Jurassic, for the Jura Mountains in Switzerland.

The discovery of radioactivity late in the nineteenth century provided geologists with both a chronometric scale for geological time and the mechanism by which to determine the absolute age of the earth. Radioactive decay is the process whereby some elements that occur naturally within the earth undergo a spontaneous change such that some of the atoms of the radioactive element change into atoms of another element. The rate at which the change takes place, the half-life, can be measured and, once measured, serves as a geological clock.

Radiometric dating, the use of radioactive elements to determine the age of rocks, has limitations. Only certain materials, be they rocks or fossils, are suitable for analysis. Igneous rocks are the best suited for radiometric dating, because as the rocks solidify from magma, radioactive elements become incorporated into the crystal structure of minerals and immediately begin to decay. The age of the rock, the time at which it formed from the molten state, can be calculated. Ironically, only in unusual cases, such as dating the petrified trees at Ginkgo Petrified

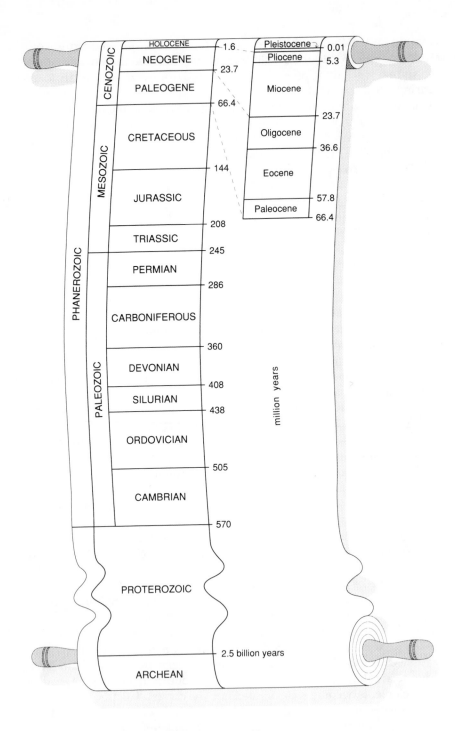

Figure 4. The geological time scale.

Forest State Park in Washington [Site 6], can igneous rocks be used to date fossils directly. Usually, dating a layer of igneous rock either above or below a fossiliferous layer allows the determination of upper or lower limits on the age of the fossils.

It is virtually impossible to radiometrically date sedimentary rocks, the rocks within which most fossils are found. In some cases, however, the fossils themselves can be dated using radioactive carbon that has been fixed in a solid state by biological activity, the biological activity in this case being analogous to the formation of crystals from magma. The limitation of radioactive carbon is that it has a short half-life and can only date fossils that are less than about 40,000 years old. The mammoth bones found at the Mammoth Site of Hot Springs, South Dakota [Site 18], for example, have been dated on the basis of radioactive carbon; the age of the bones is calculated to be 26,000±975 years (the probability is that they could be as old as 26,975 years or as young as 25,025 years).

Radiometric dating allows accurate and precise dating of various geological events in earth history; as techniques of absolute dating are refined, the absolute and relative geological time scales are integrated with increasing levels of precision. Relative dating continues to identify the order of geological events; absolute dating provides accurate dates for some of the events in the sequence. It is possible, therefore, to provide at least an estimate of the absolute age for each subdivision of relative time and the duration of each.

James Hutton in the seventeenth century wrote of geological time: "no vestige of a beginning, — no prospect of an end." [4] Deep time, immeasureable. Finally, in the early years of the twentieth century, with the discovery of radioactivity, deep time became measureable.

Biological Evolution

Biological evolution, proclaimed to the world in 1859 when Charles Darwin published *On the Origin of Species by Means of Natural Selection, or the Preservation of Favoured Races in the Struggle for Life*, is perhaps the most controversial idea to emerge in the entire history of human thought. It is particularly powerful, because it describes the elegant and coherent unity within the biological world and links the biological world irrevocably with the physical world. It has proved to be a disturbing idea, however, because it removes humanity from the apex of life's scale of beings and places her on a small twig on the margin of a vast and diverse tree of life.

---

[4]James Hutton, 1788, Theory of the Earth. *Transactions of the Royal Society of Edinburgh*, p. 304.

Long before Charles Darwin published *The Origin of Species*, ideas about evolution were widespread. Among the first to argue that species evolve and become extinct was Robert Hooke, a polymath in the late seventeenth century world of science. He, like Steno before him, recognized that fossils, petrifactions as he called them, were the remains of plants and animals. He argued that some species are modified and others become extinct in response to profound changes in the landscape. He recognized, for example, that landmasses have been repeatedly flooded during geological time and that the floods changed the physical landscape. The physical changes, he argued, produced changes in plant and animal species. Other natural historians, both in England and in continental Europe, developed similar evolutionary ideas throughout the eighteenth and early nineteenth centuries. Even Charles Darwin's grandfather, Erasmus Darwin, wrote poetically about evolution of life.

Charles Darwin was a keen observer of nature, his enthusiasm for both geology and biology nurtured during his days as a student in Cambridge. His five-year voyage around the world on the *Beagle* (1831–1836) was ideal training for a natural historian, and during the course of that voyage, Charles Darwin began to consider evolution to account for the diversity he saw in nature. Upon his return to England, he quickly established himself as a leading scientist of the day and began to consider what he called the transmutation of species. He began his first notebook on the subject in July, 1837, shortly after his return on the *Beagle*. For twenty years he contemplated and studied, developing his ideas on evolution. He was reluctant, however, to publish his ideas, sensitive to the public outrage he was sure would follow. Finally, in 1858, he was compelled to act, motivated by a manuscript sent to him by Alfred Russel Wallace from southeast Asia. Wallace had independently arrived at the concept of natural selection for the origin of species and wanted Darwin's opinion of his work. Within weeks, Darwin's and Wallace's ideas were presented jointly at the Linnean Society in London, and late in 1859, *The Origin of Species* appeared in print. There was no turning back for Darwin then; the world has not been the same since.

History has not been kind to Wallace, his role in the formulation of the concept of natural selection for the origin of species largely forgotten. Full title accrues to Charles Darwin largely because he published so promptly *The Origin of Species*, a work which marshalled encyclopedic observations to support his conclusions. Darwin's argument for biological evolution is grand in its simplicity. Although he was not alone among natural historians in recognizing that organisms are not static, that they have changed through time and are different from place to place, his contribution is that he presented a theory that explains the changes in organisms.

Darwin's logical argument for evolution is based on four main and cumulative points:

1) Darwin observed that organisms of any species produce more offspring than are needed to maintain the population and more than can possibly survive.

2) Darwin recognized that in the natural world individuals must compete with one another for such necessities for survival as food, shelter, and living space.

3) Darwin also observed that organisms vary in nature, that individuals within a species differ in any number of characteristics (such as overall size or color of ornamental plumage in some birds). Furthermore, this variation is heritable, passed from parent to offspring in the reproductive process.

4) Darwin argued that the presence of certain variations in some individuals gives them an advantage, that they are better able to survive in any particular environment. These are the individuals that, over the long term, are most likely to live and reproduce. The favorable variations are then passed on to their offspring. Darwin called this mechanism natural selection because the mechanism in the natural world is analogous to artificial selection as practiced by plant and animal breeders.

By means of natural selection, favorable variations are accumulated by succeeding generations, and new species arise.

Species are the basic groupings into which all life on earth is organized; all individuals on earth belong to one and only one of these groupings, each grouping is distinct from all others. Since species arise by descent with modification, the biological world can be organized into an enormous kinship tree that illustrates the relationships, both similarities and differences, among species. The resulting classification is hierarchical and is analogous to human family genealogies.

The modern system of classification contains seven levels, with the species the basic unit. Very similar, very closely related species are grouped together in genera. Thus, the coyote (*Canis latrans*) and the wolf (*Canis lupus*) are recognized as closely related species, siblings in the human family analogy. These species and their cousins, foxes such as the red fox (*Vulpes vulpes*) and the kit fox (*Vulpes macrotis*), are members of the Family Canidae. Members of the dog family are distinct from members of the cat family (Family Felidae—cougars, bobcat, lynx, etc.) and the bear family (Family Ursidae—black bear, grizzly, polar bear, etc.), but the similarities among the families are strong enough to indicate a distant relationship, one that is acknowledged by classifying them together in the Order Carnivora. Other mammals are similarly classified to the order level and united in the Class Mammalia, because all share certain reproductive and biochemical strategies. Equivalent groups are the classes Aves (birds), Reptilia (reptiles), Amphibia (amphibians), and Pisces (fishes). That they are distantly related can be seen in the common possession of a vertebral column; hence, they are classified—along with a few ancestral organisms that possess the precursor to a vertebral column (a notochord)—within the Phylum Chordata.

Phyla are the fundamental body plans of life, each phylum defined by a basic anatomical structure that allows the organisms it comprises to make a living in a distinct way. Animal life (Kingdom Animalia) as we know it today is made up of six distinct body plans; chordates have a distinct body plan from that of arthropods, mollusks, annelids, corals, or sponges. The fossil record shows, however, that animal life has not always been limited to six body plans. Fossils from the Burgess Shale in British Columbia [Site 4] reveal many unusual body plans unlike any known today.

From an evolutionary perspective, it is significant that, at different times in the history of the earth, life forms are organized at different levels. At the present time, for example, diversity—the number of distinct species within a body plan—is high, but the disparity—the number of different body plans—is low. Early in the history of life, the organization was different; the disparity was higher, the diversity was lower.[5]

At the highest level of classification, all life on earth, living and extinct, is grouped into one of five kingdoms: Animalia, Plantae, Fungi, Protoctista, and Monera. The division reflects the ultimate distinctness of life, which is based on unique biochemistry and is manifest only indirectly in the fossil record.

Darwin recognized the central role that fossils would play in documenting evolution. As soon as they were recognized to be the remains of once-living organisms, fossils begged questions of evolution; their very presence and the diversity among them, the proof of change over time in the biological world (Figure 5). Darwin explained the mechanism by which evolution takes place, the mechanism of natural selection; modern genetics identifies mutations as the ultimate source of heritable variation.

The theory of evolution explains how new species arise. By extension, evolution must also explain how species become extinct. Origin and extinction are, after all, opposite sides of the same phenomenon. Just as birth and death define the life of an organism, so do origin and extinction define the life of a species. Natural selection is a creative force acting upon individuals within species and upon the different species within any environment.

Extinction, too, is a creative process; it clears the way, opening physical spaces into which new species might evolve. Paleontologists recognize three different intensities of extinction in the fossil record. A low level of extinction forms the normal background in the fossil record, origin and extinction of species keeping pace. Then there are short

---

[5]I use the terms disparity and diversity in the sense articulated by Stephen Jay Gould, 1989, *Wonderful Life*, p. 49, and other authors (in more technical papers).

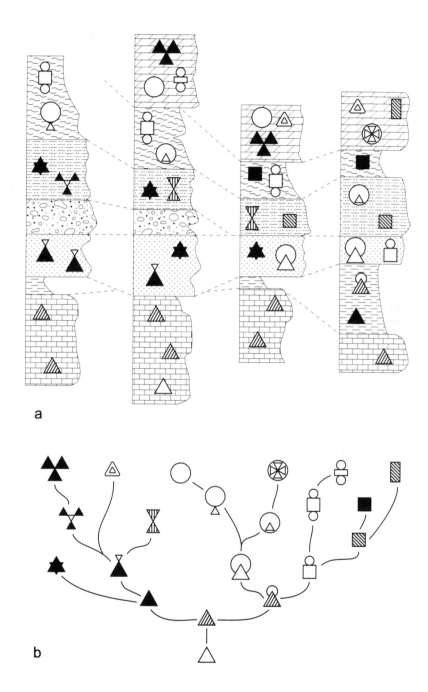

Figure 5. The role of fossils in documenting evolution. In **a**, layered rocks are correlated on the basis of the fossils they contain; in **b**, the fossils are organized into a family tree to show their inferred evolutionary relationships.

27

periods of time when the intensity of extinction is very high, and many species suffer extinction simultaneously. These are times of mass extinction, spectacular phenomena in the fossil record characterized by the elimination of a large percentage of the biota over a very short segment of geological time. Between times of mass extinction, the steady-state level of evolution and extinction was punctuated periodically by small extinction events.

The fossil record as we know it verifies six episodes of mass extinction in the history of life on earth: 1) late in Precambrian time, 2) at the end of the Ordovician, 3) late in Devonian time, 4) at the end of the Permian, 5) at the end of the Triassic, and 6) at the end of Cretaceous time. The nature of life was irrevocably altered by each mass extinction; the reigning order was eliminated and completely new assemblages of organisms evolved to fill the void.

Paleontologists recognize that the intensity of origin of new species in the fossil record is directly related to the intensities of extinctions. There is a background level of origins that is analogous to the background level of extinctions. What is significant is that each episode of higher-than-background extinction is followed, either immediately or after a lag period, by an episode of higher-than-background origins; each mass extinction is followed by adaptive radiation, a burst in the number of new species that are suddenly present in the fossil record. Mass extinctions punctuate the fossil record; adaptive radiations are dramatic revolutions in the history of life.

The search for the causes of mass extinctions has been one of the most elusive in paleontology. The series of events that surround each mass extinction have been documented; it is quite another problem to ascribe a cause and effect relationship between such historical events and the mass extinction. For example, there is no doubt that mountain building was active in Late Cretaceous time especially on the west coast of North America, that the continents of the world were becoming drier as epicontinental seas withdrew from inland regions, and that climate world-wide was becoming cooler as the continents drifted toward the poles. One can cite these gradual changes to explain the steady-state changes in the faunas and floras of the Late Cretaceous and to document ecosystems under stress. Such gradual changes, however, are insufficient to explain sudden, dramatic mass extinction of many organisms world-wide, among them the dinosaurs, for though the changes describe the historical context they do not define cause and effect relationships.

Some paleontologists argue that there are global patterns of extinction through geological time and that those patterns are independent of tectonic processes and climatic pressures. The global patterns suggest that catastrophic extra-terrestrial events, such as meteorite impacts, are the agents of mass extinction.

The discovery of anomalously high levels of iridium in rocks that formed at the very end of the Cretaceous Period offers a solution to the problem of explaining mass extinction because the iridium is derived from an extra-terrestrial source, be it asteroid, meteorite, or comet. The impact of a large extra-terrestrial body with the earth is the only known way to introduce a sudden catastrophe of sufficient magnitude to upset the steady-state balance of the earth and its biosphere and, thereby, precipitate mass extinction on a global scale. There is even evidence that extinction events may be repetitive and cyclic, that the earth may be periodically subjected to showers of extra-terrestrial bodies.[6]

The study of the twin phenomena of mass extinction and adaptive radiation is extending the multi-disciplinary scope of paleontology. Today, the strong argument in support of an accidental or extra-terrestrial cause for mass extinction and, perhaps, for intermediate-level extinction events suggests that astronomy, too, intersects with biology and geology within the scope of paleontology.

More than one hundred and thirty years have passed since Darwin first published *The Origin of Species.* During that time, scientific understanding of evolution has itself evolved, having been modified by new data and increasing knowledge. Darwin's broad vision, nonetheless, continues to characterize biological evolution. Paleontologists debate the fine points of evolution; they disagree on rate and process, on the role of episodic events, on hierarchy and continuity. But they do not deny evolution.

---

[6]See David M. Raup, 1986, *The Nemesis Affair,* for an in-depth review of the Late Cretaceous extinction event and the implications of extra-terrestrial impacts for mass extinctions.

The History

Origins: Land and Life

The origin of the earth, the origin of life on earth: there was a time when these were discrete topics of scientific investigation, when geological and biological processes were thought to operate independently of each other. That is no longer the view. Today we have evidence of life in some of the oldest rocks on the surface of the earth. We have evidence that the presence of life altered geological processes, that the origin and evolution of life is an integral component of the evolution of the earth.

The earth formed about 4.6 billion years ago as part of the solar system from clouds of solar dust that rotated about the sun. By human standards, the earth was not a comfortable or inviting place during the early stages of its formation; rather, it was a mass of extremely hot, molten rock without the atmosphere and the liquid water that characterize it today. As the primitive earth slowly cooled, the gases locked inside the molten rock began to escape, and the atmosphere began to form. The process of degassing continues to this day as volcanoes erupt, expelling gases and molten rock. The early atmosphere was composed of volcanic gases, the same gases that volcanoes release at the present time: water vapor, hydrogen, hydrogen sulfide, hydrogen chloride, carbon dioxide, methane, and others. Oxygen was absent!

About 4.0 billion years ago, when the earth had cooled sufficiently, molten rock began to solidify and form interlocking chunks of solid crust. As the temperature on the surface dropped below 100 degrees Celsius (212 degrees Fahrenheit), water vapor in the atmosphere precipitated. Rains fell and oceans formed. In these early oceans, in the absence of atmospheric oxygen, life first appeared.

Among the earliest forms of life on earth were primitive bacteria and cyanobacteria (the so-called blue-green algae), acellular or very simple single-celled organisms. They manufactured their own food using energy derived from chemical reactions (chemosynthesis) or, as did the cyanobacteria, by photosynthesis using energy derived from sunlight.

Organisms dependent on chemosynthesis could live on the ocean floor in deep water, but organisms dependent on sunlight must live within the photic zone, that critical layer of water at the surface through which light will penetrate. During photosynthesis, organisms absorb carbon dioxide from the atmosphere and, using sunlight, combine it with nutrients present in the ocean to produce food; in the process, free oxygen is generated as a waste product. The oxygen was an introduced substance, toxic to the living organisms found on primordial earth; the early cyanobacteria were polluting the environment.

Primeval organisms were microscopic. Could they possibly have been preserved? Can they even be recognized in the ancient and strongly altered rocks of the Precambrian? Amazingly enough, yes, even the simplest of all life on earth, living 3.5 billion years ago, left structures as evidence of their existence. The structures are stromatolites, layered mounds of mud and organic material; the organic material, the remains of bacteria and the filaments of algae. These fossils provide the strongest evidence of the oldest life on earth.

Stromatolites became abundant about 2.5 billion years ago, in Proterozoic time. A sophisticated assemblage of single-celled organisms banded together to form complex ecological communities, broadly analogous to the complex modern communities that we call reefs. The organisms that made up the stromatolite communities were significant forms of life throughout the Proterozoic; their remains are readily visible in many rocks, including some in Waterton-Glacier International Peace Park in Alberta and Montana [Site 5].

Life may well have been simple in structure, but it was widespread in all the oceans of the ancient world. Even so, life was organized very differently in the Proterozoic than it is today. The number of different kinds of body plans—the disparity—was high; the number of distinct organisms within each kind of body plan—the diversity—was low.

Meanwhile, the earth continued to cool; over time, continents and distinct ocean basins formed. Evidence from rocks of Early Proterozoic age indicates that, as early as 2.5 billion years ago, small continental landmasses were present and were separated by discrete ocean floors. It is no coincidence that stromatolites had become abundant more-or-less simultaneously as the first small continents had developed. The level of development of the physical earth was significant, because the presence of continents meant that there were now shallow-water ocean environments surrounding the continents. Such environments were critical for photosynthetic organisms dependent on inorganic nutrients derived from the land and on adequate sunlight for the energy to produce their own food. In the shallow waters surrounding the small continents, life proliferated.

The physical processes of the earth assumed a modern aspect with the development of the first small continents. For the last 2.5 billion years, these processes, described collectively as plate tectonics, have shaped the earth. The crust has been made up entirely of rigid plates in constant motion relative to one another (Figure 6). Some plates or portions of plates are oceanic crust, a thin layer of dense rock (basalt) that forms the ocean floor. Other plates are thick accumulations of lighter rocks and form the continents. Plates are kept in motion by the heat dissipating from within the interior of the earth. Where plates fracture and move away from each other, molten rock spills through the cracks and solidifies to form new oceanic crust. Where they collide, new mountain chains are formed; old oceanic crust is forced beneath the leading edge of the lighter plate with which it is in contact (along a subduction zone), driven deep into the interior of the earth and destroyed. If two continental landmasses come into contact, neither is dense enough to slip beneath the other; instead, the light rocks are crumpled and piled up to form massive, broad and high mountains.

The role of stromatolite communities in modifying the physical world, in particular the early atmosphere, cannot be overstated. Organisms extracted carbon dioxide from the atmosphere and precipitated it in the form of calcium carbonate (limestone) rock on the sea floor, simultaneously releasing free oxygen. Free oxygen was thus introduced into the atmosphere. Equally important, however, was the removal of carbon dioxide. The latter is a greenhouse gas; its depletion from the atmosphere insured a temperate climate for the earth. To this day, organisms continue to extract carbon dioxide and turn it into organic rocks, thereby maintaining the delicate balance of gases in the atmosphere upon which life now depends. It should also be pointed out that, in the process of removing carbon dioxide from the atmosphere, organisms simultaneously remove calcium from solution in the oceans. Without a mechanism present for fixing calcium in solid form, the concentration of calcium ions in the oceans would quickly rise to toxic levels.

The oxygen produced by the early forms of life slowly changed the chemistry of the earth. At first, all the free oxygen that organisms generated was immediately used in chemical reactions (rusting, for example, turns pure iron into iron oxide). Eventually, about 2.0 billion years ago, all the chemical reactions for using oxygen were saturated, and free oxygen began to accumulate in the atmosphere. Some, however, was altered to ozone; as the amount of free oxygen increased, an ozone layer developed in the upper atmosphere. It offered protection to organisms from harmful ultraviolet radiation, a protection that had previously been afforded either by submersion beneath an adequate depth of water or by smog-like gases in the atmosphere (among them the carbon dioxide and methane, which were being removing biologically from the atmosphere).

Free oxygen in the atmosphere altered the chemistry of life. Aerobic biochemical reactions are much more efficient than anaerobic ones at converting the food an organism burns into the energy it requires to

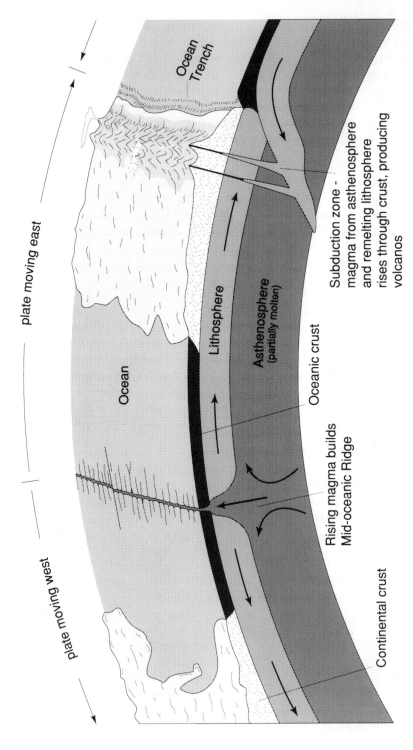

Figure 6. Plate tectonics. The surface of the earth is made up of rigid plates that are in constant motion relative to one another. The interactions between plates at their boundaries shape the surface of the earth.

grow and reproduce. As the level of free oxygen increased so did the efficiency and complexity of aerobic biochemical pathways.

As early as about 2.0 billion years ago, the level of internal organization of life forms crossed a major threshold; the internal structure of single living cells became organized into discrete subunits, the genetic material in particular becoming concentrated in a nucleus. This development may well mark the beginning of aerobic organisms. The earth's ecosystem was becoming increasingly complex, its elements increasingly interdependent.

The second threshold in the organization of biological structure was crossed as groups of cells became united to build simple, multicellular organisms. The time of the first appearance of animals, those organisms that consume organic material produced by other organisms (primarily through photosynthesis), is unresolved. It may be that the first animals are to be found among the first aerobic, single-celled organisms; it seems certain, however, that animals evolved, at the very latest, among the first simple, multicellular organisms. Regardless, late in the Proterozoic, some 650 million years ago, complex, multicellular, soft-bodied organisms appeared—both plants and animals, producers and consumers. They were first found as fossils in the Ediacara Hills of Australia and are now found preserved in rocks all over the world; the age of the fossils, Ediacarian, is named for the site of first discovery

With the evolution of multicellular organisms, the stage was set for the appearance of organisms characterized by large size, anatomical complexity, and hard parts. Physical events continued to build and shape the continents and to constrain the course of biological evolution. In North America, as in all the continents, the fossil record must be interpreted in the light of the geological events that built the continent itself.

The Panorama of Life

Apparently suddenly, all over the world in rocks of Early Cambrian age, fossils abound. Apparently suddenly, the fossil record is rich, documenting phenomenal biological diversity and the rise and fall of organisms, illustrating catastrophic events of global proportion, and preserving both small and mighty for posterity. For many decades, Cambrian fossils were the earliest evidence of life, and the so-called Cambrian Explosion was thought to document the origin of life on earth. With the confirmation that life is almost as old as the earth itself, the Cambrian Explosion was relegated to marking the evolution of organisms with hard parts. Over the last thirty years even that distinction has blurred, for it has been amply shown that the evolution of skeletons occurred some 30 million years before the Cambrian Explosion. Obviously, a profound restructuring of life occurred in the latest Precambrian, the Cambrian Explosion marking but the beginning of the familiar fossil record.

The fossil record is an epic story of the different groups of organisms that succeeded one another through geological time. It proves that life on earth is dynamic, constantly evolving in the face of adversity; it proves that life is persistent and successful, its variety insuring that no catastrophe has been able to eliminate all organisms. Inevitably, some species survived, and these quickly repopulated the world.

Late Precambrian Restructuring

Life began in the oceans when the world was young. Over the eons, organisms effected changes on the planet and then themselves underwent profound restructuring in response to the conditions life created. Late Precambrian time in particular was a great watershed in the history of life.

Late in Precambrian time, some 650 million years ago, the Ediacarian fauna, typified by fossils from the Ediacara Hills of Australia, existed round the world. The organisms, preserved as impressions on what was the soft and sandy sediment of the ocean floor, were large, complex, multicellular, and entirely soft-bodied.

The multicellular level of organization afforded tremendous potential for biological complexity. For the first time in the history of life, it became possible for organisms to achieve large size. But size, too, has its limitations. For organisms, the problem is one of scaling; as the size of an object is increased, the amount of surface area increases less rapidly than does the volume. Many organic functions, among them respiration and digestion, depend on adequate surface area. In order for organisms to achieve large size, therefore, there must be a way for them to maximize surface area relative to volume. One such way evolved in Ediacarian time, large organisms shaped like pancakes and ribbon. Ediacarian-style of life persisted for 50 or 75 million years, then abruptly disappeared. It is the first mass extinction of life for which there is evidence from the fossil record.

The appearance of shelly organisms marks the recovery of life after the Ediacarian mass extinction, an adaptive radiation of life forms characterized by a new and distinctive mechanism to solve the problem of decreasing surface area with increasing size. The new organisms retained a bulbous shape as they grew larger but increased surface area by developing internal organs from infoldings of the surface. The modern fossil record, beginning in the Paleozoic, illustrates the way in which biological evolution has been able to elaborate on this basic design.

The shelly organisms constituted but a small proportion of the Paleozoic radiation. They illustrate a significant step in evolutionary opportunism—they were able to take a biochemical mechanism, the removal of toxic calcium ions from cellular fluids by precipitation in the form of solid calcium carbonate, and adapt it for direct functional purposes. Thus was formed a protective shell and a support structure for soft tissues. These hard parts are the essentials of the fossil record.

Ancient Life

The Paleozoic, the time of ancient life, began about 600 million years ago with the first appearance of skeletal fossils. The seas teemed with various complex and abundant life forms. It was an exciting time in the history of life, a time that witnessed the dominance of invertebrates in the seas, the rise and diversification of vertebrates, and eventually the colonization of land by both plants and animals.

Algae in the oceans remained the primary producers of food in the Paleozoic ecosystem. They photosynthesized organic molecules using the energy of sunlight to transform carbon dioxide and water into sugars. Animals, unable to manufacture their own food, were the primary consumers; grazers and scavengers, they consumed the organic molecules produced by algae.

Ancient life, for the first 30 million years, was represented by the so-called Tommotian fossils, widely known but poorly understood tiny shells. These tube-, coil-, or cup-shaped fossils represent biological readjustment to the devastating Ediacarian mass extinction. Tommotian time was succeeded by the classical Cambrian Explosion and the first trilobites.

By Paleozoic time, the physical world was taking on a modern configuration: continental landmasses and extensive ocean basins were well established. The continents were situated straddling the equator. Sea level began to rise, and the oceans of the world flooded over the continents (Figure 7). The physical development had significant implications for organisms, because as increasingly larger portions of the land were flooded, the total surface area of shallow-water environments was greatly increased. Warm, tropical epicontinental seas are rich in nutrients, oxygen, and sunlight and are, thus, particularly conducive environments for organisms. A vast area of the world was being made habitable, and within it were many environments into which marine organisms diversified.

The areas of the oceans that are the richest for biological growth occur in tropical regions, along the shallow margins of continents and the margins of deep, inland basins that are flooded by epicontinental seas. The topography of the sea floor creates a marked transition from the deep water of oceans to shallow conditions along the shoreline. Within that zone, where the wave energy is highest, the water is well oxygenated and richly supplied with nutrients; the two conditions are critical for life to flourish. On the sea floor within the photic zone, photo-

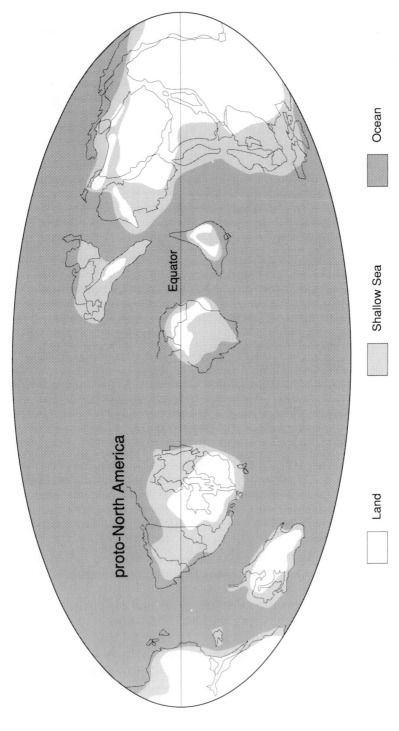

Figure 7. Paleogeography of early Paleozoic time. Modern continents are not recognizable, but continental landmasses with shallow seas encroaching on land were well established.

Land   Shallow Sea   Ocean

proto-North America

Equator

synthetic organisms proliferate. Warm temperatures insure that organisms thrive. The combination of conditions is ideal for complex ecological communities to develop and thrive.

Ideal environmental conditions are meaningless if there are no extant organisms to exploit them. In particular, if there exist organisms capable of secreting calcium carbonate skeletons and simultaneously cementing those skeletons together to form a solid, three-dimensional structure that rises up off the sea floor, a distinctive ecological community will form—the biological reef.[7]

In Cambrian time, physical and biological conditions conjoined to produce the Cambrian Explosion. Trilobites are the most conspicuous fossils of the Cambrian Period. Readily recognized on the basis of the tri-lobed skeleton, they are ancient arthropods with segmented bodies and jointed limbs (Figure 8). Some were free-floating, others were active swimmers, but most lived on the sea bottom extracting organic nutrients from the sediments.

We see the first of nature's experiments at reef building, for a brief time, early in the Cambrian. These were true biological reefs, distinguished from the stromatolites of earlier time by the presence of archaeocyathids, organisms that actively secreted and cemented calcium carbonate skeletons. The archaeocyathids were sponge-like suspension feeders that built vase-shaped skeletons (Figure 8) and obtained their food by filtering out small organic particles suspended in the surrounding water. Stromatolite-building algae contributed to reef building; so did animals such as brachiopods, mollusks, and echinoderms that would become significant in later time. The archaeocyathids survived for only a short time in the Cambrian; when they became extinct, biological reefs disappeared for a time from the fossil record.

The Cambrian Period is amazing for its extraordinary disparity of life. During that time, life was characterized by the presence of many different kinds of body plans, some body plans so unlike the ones that characterize modern animals that the organisms were truly bizarre. One special fauna, from the Burgess Shale in British Columbia [Site 4], preserves an assemblage of soft-bodied organisms and reveals the unimagined variety of body plans extant in the Cambrian. Equally important, it

---

[7]The term "reef" is a confusing one and has many meanings. It originally referred to any structure on the sea floor upon which a ship might flounder. In paleontological parlance, a biological reef is a structure built by organisms as they secrete their skeletons that stands in positive relief on the sea floor during the life-time of the organisms. Many ancient organic structures (for example, the Permian Capitan Reef in Guadalupe Mountains National Park of west Texas [Site 30]) exhibit three-dimensional relief as they are presently exposed in rocks, relief that is an artifact of burial and preservation, and they are called reefs. They are not, however, biological reefs unless they were built by reef-building organisms and had positive relief at the time they were built.

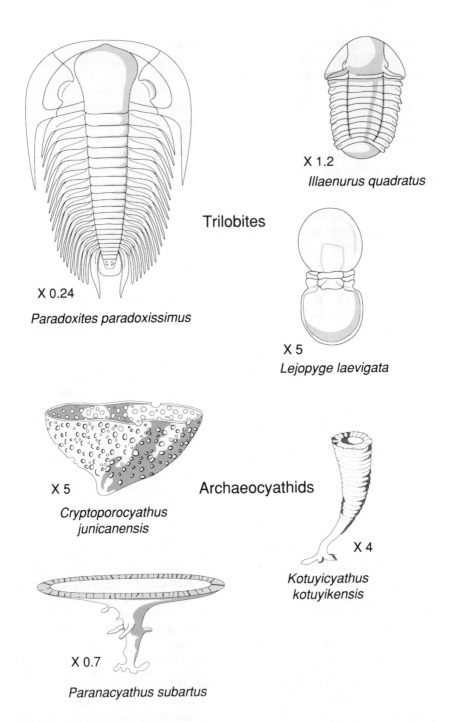

Trilobites

X 1.2
Illaenurus quadratus

X 0.24
Paradoxites paradoxissimus

X 5
Lejopyge laevigata

X 5
Cryptoporocyathus junicanensis

Archaeocyathids

X 4
Kotuyicyathus kotuyikensis

X 0.7
Paranacyathus subartus

Figure 8. Some Early Cambrian marine invertebrates: trilobites at top and the vase-shaped skeletons of archaeocyathids at bottom (modified from *Treatise on Invertebrate Paleontology*).

reveals that the number of species, the diversity, within each body plan was low; most body plans were represented by a single species, this in contrast to the large number of species within body plans today (for example, living vertebrates are represented by hundreds of species, insects by thousands). In other words, the complexity of life in the Cambrian, marked by high disparity but low diversity, was very different than it is today. It is important to note, in addition, that a completely modern ecological association existed: the identification from the Burgess Shale of the oldest known true predators confirms that the food pyramid was complete, comprising primary producers at the base and ultimate consumers at the apex.

Only a few of the many body plans of Cambrian time survived extinction about 550 million years ago. Disparity was eliminated; diversification became the mechanism for increasing evolutionary complexity. In a way, the Cambrian Explosion did not really happen until the Ordovician. True, suddenly in Cambrian rocks, fossils are abundant; and, true, life was abundant in Cambrian oceans; but the complexity was different. In the Cambrian, there were abundant individuals representing a great disparity of body plans; in contrast, the explosion of life in the Ordovician was an explosion in diversity within a limited number of body plans. The organization of life as we know it had become established.

The end of Cambrian time is marked by the extinction of most of the species of trilobites. Although some forms persisted for millions of years, trilobites were never again as common or important.

Ordovician time is defined by adaptive radiation, the dramatic increase in diversity of many groups of marine invertebrates. Among them were brachiopods, graptolites, conodonts, rugose corals, crinoids. No one group dominated. Primitive jawless fish, the first known vertebrates, probably evolved in the Cambrian, but they first appear in the fossil record in Ordovician time. Bryozoans, stromatoporoids, and tabulate corals gradually became more common, and by Middle Ordovician time these organisms began nature's second experiment in reef building. They formed the so-called tabulate-stromatoporoid reef community, the successful reef builders in the world's oceans for about 100 million years, through Silurian and Devonian time. Sea level was high, and the interiors of the continents were almost completely flooded. Life was rich in such extensive epicontinental seas. The fossils at Stonewall Quarry Park in Manitoba [Site 1] illustrate both the extent of Ordovician seas and the diversity of organisms that lived in them (Figure 9).

Profound extinction occurred at the end of the Ordovician Period; many species of many groups of animals disappeared. It is the second of six mass extinctions that decimated life on earth. The causes are poorly understood. Tectonic processes may have played a role. The continents of the world, over time, had drifted away from the tropical positions they had occupied in the early Paleozoic and toward the poles. The continents

Figure 9. Diversity of life in Late Ordovician seas. The diorama at the Saskatchewan Museum of Natural History reconstructs marine life in the Williston Basin.

of the southern hemisphere had become fused to form a single continent, Gondwana; by the end of the Ordovician, Gondwana was situated over the South Pole. Glaciers developed.

Did a world-wide drop in temperature precipitate mass extinction? Perhaps. Is it possible that the glaciers stored sufficient water to drop sea level to a point intolerable to many organisms? Perhaps, but in North America the evidence is contrary. The Ordovician epicontinental seas persisted well into Silurian time, until well after the mass extinction; only then did the seas rapidly drain away and afterwards gradually flood back again.

The Ordovician mass extinction conforms to perceived global patterns of extinction over geological time, patterns that argue for catastrophic extra-terrestrial events as causative agents. The evidence of a meteorite or comet impact remains equivocal, but the fossil evidence that mass extinction occurred cannot be denied.

Life in the oceans recovered quickly after the Ordovician mass extinction. Many species of the major groups of organisms had disappeared, but those that survived rapidly evolved to produce a diversity of new species. Among them, in the Silurian and Devonian, were new brachiopods, bivalve and gastropod mollusks, bryozoans, crinoids, and graptolites. Marine life, overall, was not very different: the new species that evolved to occupy the habitats that the extinct species had once occupied were not unlike their precursors.

The reef-building experiment that began in the Ordovician was notably unaffected by mass extinction and became enormously success-ful in the Silurian and Devonian periods. The tabulate-stromatoporoid community, tabulate corals and stromatoporoids with associated colonial rugose corals and bryozoans, built massive structures that rose up off the sea floor to form barrier and pinnacle reefs analogous to those that grow in modern oceans.

The reefs grew in areas of deeper waters in the inland seas. With time, as the reefs grew more and more massive, they began to alter the environments in which they lived by restricting the circulation of water within the seas. Evaporation had a startling effect in the restricted areas: the salts in the water became increasingly concentrated, and as more water was removed, the salts precipitated to form evaporites such as gypsum and potash.

The waters surrounding the reefs were rich and supported a myriad of organisms; among them were new free-swimming animals which emerged to become the true predators in the oceans. The first ammon-oids, the first jawed fish, and eurypterids (Figure 10): the first known hunters since extinction removed the soft-bodied predators of the Cam-brian seas. The ammonoids would become important and diverse invertebrates in late Paleozoic and Mesozoic oceans. Fish would not only be numerous and successful in the oceans up to the present time, but among the various groups of fish there would be one that would even-tually make the transition to amphibian and to terrestrial habitats. The eurypterids, a group of extinct arthropods closely related to the living horseshoe crab, were formidable in Paleozoic seas, especially in the hypersaline waters of the Silurian and Devonian epicontinental seas, in the lagoons and estuaries that formed as water circulation was restricted.

Life was apparently confined to the world's oceans until the middle of the Paleozoic; not until Silurian and Devonian time is there any evi-dence that life was present on land. The terrestrial environment was alien to forms of life whose origins were aquatic. The requirements of life on land are very different than requirements in water: dehydration is a constant threat, even if a protective covering is present to minimize water loss, because the supply of water is not always constant; oxygen must be extracted from air rather than water; reproduction depends on adequate moisture. The physical properties of air are very different from those of water, so organisms must have structures for support, animals mechanisms for locomotion. In spite of such obstacles, plants, verte-brates, and some arthropods, including the ancestors of insects, spearheaded the colonization of land.

It seems certain that the movement of life onto land was controlled by the level of development of the atmosphere, that only when the con-

Figure 10. Reconstruction of *Eurypterus lacustris* of Late Silurian time (Smithsonian Institution Photograph No. SI–88–16880).

centration of free oxygen was sufficiently high could the high metabolic rates necessary for life on land be sustained. The terrestrial environment then offered distinct advantages to those organisms able to overcome the physical constraints to life on land.

Sunlight is the advantage that terrestrial environments offer. Water, while it affords protection from radiation, also dissipates much sunlight. Only a thin layer of water at the surface is lit; for that reason, most life in the oceans is restricted to the uppermost layer, the photic zone, where photosynthesis is possible. Air, in contrast, does not impede sunlight. The greater access to sunlight on land was of tremendous advantage to plants; very quickly in Silurian and Devonian time, a complex terrestrial ecosystem evolved, based on the success of land plants. The terrestrial world also offered empty space, not an insignificant factor when compared to the densely populated seas of the mid-Paleozoic.

The first putative terrestrial organisms are plants. The oldest fossils of terrestrial plants are Middle Silurian in age, but there is evidence that some plants began the transition to land as early as the Late Ordovician. The first known animals on land are millipedes and other arthropods of

Late Silurian age, and by Early Devonian time, both plants and arthropods were well established in terrestrial areas. Later in the Devonian, insects first appeared, and vertebrates made the transition to land.

Every once in a while, the fossil record preserves examples of transitional organisms, forms that span discontinuities in evolutionary history. One such form is *Ichthyostega*. It is one of a group of amphibians, the ichthyostegids, that documents the evolutionary transition of vertebrates from life in the ocean to life on land. A specialized branch of Paleozoic fish (rhipidistean fish) underwent a series of anatomical modifications that eventually took fish out of the water and put amphibians such as *Ichthyostega* on land. Amphibians themselves bridge the biological discontinuity between a totally aquatic lifestyle and a wholly terrestrial one. They are terrestrial animals that are tied to the water by their reproductive cycle. *Ichthyostega* is clearly amphibian, with limbs for walking on land. It had shoulder and pelvic bones that were weight-bearing structures. The bones of the vertebral column were fused and interlocking for added strength. Yet, *Ichthyostega* retained characteristics of its piscine past: a powerful fish-like tail for swimming, gills for breathing underwater. It was still capable of functioning in the water with ease.

The early amphibians were large animals, up to one meter long, active predators with sharp fangs and large, gaping mouths. With only the likes of millipedes and insects as prey on land, they undoubtedly fed on animals that lived in streams, ponds, and swamps. Capable of life on land, ichthyostegids were constrained by their ancestry, by their evolutionary history, to return to the water. There they obtained their food, and there they reproduced.

The first terrestrial vertebrates were still part of the aquatic food pyramid. It is one of the curious ironies of evolution that the first vertebrates to make the transition to land, being made green and habitable by the earlier evolution of land plants, should be carnivores, not herbivores. Herbivory among land vertebrates is a secondary phenomenon.

Late in the Devonian, mass extinction once again decimated plant and animal life in the oceans. Tabulate corals and stromatoporoids became extinct; reefs disappeared. The groups of organisms that lived in tropical seas suffered the greatest losses. Once again, the causes of mass extinction are not clearly understood, but the sudden devastation is consistent with arguments for catastrophic or extra-terrestrial events.

Marine life recovered quickly and, in the Carboniferous and Permian periods, was significantly changed in only one way. There were no reef-building organisms, and no true biological reefs are known from this time. Ironically, the most famous fossil so-called reef-complex known in the world is the Capitan Reef from the Permian of Texas [Site 30]; not a biological reef, it warrants the title reef because a three-dimensional structure stands in relief today. The shallow oceans of the world were occupied by such organisms as crinoids, rugose corals, and bryozoans,

45

suspension feeders that held themselves erect off the ocean floor, and by brachiopods and gastropods that lived in and on the ocean floor. In the upper layers of the water, microscopic plankton flourished. The crinoids were especially amazing because they covered vast areas of the shallow seas with dense thickets so reminiscent of pastoral settings that they are called crinoid meadows, their remains forming extensive sheets of limestone rock visible in mountain outcrops such as those at Minnetonka Cave in the Bear Mountains of Idaho [Site 14].

The continents of the northern hemisphere during the late Paleozoic were being assembled to form Laurasia, the giant northern continent. In the process, landmasses collided; mountains were built. The most spectacular of the mountain chains was the Hercynian Mountain Chain, a colossal range spanning Laurasia that formed when Europe and North America collided. As a consequence of mountain building, the seas gradually withdrew from the interiors of the continents, large tracts of land became subaerially exposed, and plants and animals spread across the landscape. For the first time in the fossil record, in rocks of Carboniferous age, remains of terrestrial organisms are significant components.

Beginning in the late Paleozoic, the major evolutionary innovations in the history of life took place among terrestrial rather than marine organisms. The physical world, too, had been indelibly and forever altered when life, especially plant life, moved onto the land, because plants modify weathering, minimize erosion, and thereby alter the rate and pattern of sedimentation. Physical processes on the earth were moderated for the first time.

The land became green during Carboniferous time. Plants diversified so rapidly that all the major groups but one, the flowering plants (angiosperms), were present. Many species became arborescent. Remains of fern forests are so common that the late Paleozoic is often called the age of ferns. These early plants were restricted to moist areas because they reproduced by means of spores; but, in their midst, a new group of plants, the gymnosperms, slowly emerged. They produced a seed, a plant embryo contained in a protective shell with a store of nutrients. The innovation of a seed gave plants the freedom to inhabit drier areas. The gymnosperms, naked-seed plants that hold their seeds on cones, are represented in the modern environment by conifers.

Coal! For the first time in the fossil record, the accumulated remains of plants that grew in lush swamps and bogs are preserved. In the Late Carboniferous,[8] large areas of the continents comprised extensive wet-

---

[8]The Carboniferous period is divided into two epochs, Early Carboniferous and Late Carboniferous. In the United States, the terms Mississippian and Pennsylvanian coincide with Early and Late Carboniferous usage elsewhere in the world. The Late Carboniferous is called Pennsylvanian in reference to the extensive coal deposits of Pennsylvania.

lands, a habitat so conducive to prolific plant growth that it has been given a special name, the coal swamp.

Oxygen was absent in the deeper, murky waters of the coal swamp, and organic material did not decay. Instead, the plant debris accumulated and was buried. Over time, it was altered to peat and to brown coal called lignite; then to soft, black, bituminous coal; finally, if temperatures and pressures were high enough, it was metamorphosed to anthracite. Incredible volumes of the world's total carbon supply were removed from the atmosphere and biosphere and stored deep within the earth.

Imagine the impact! If the addition of carbon dioxide to the atmosphere produces the greenhouse effect, does its removal produce the opposite? Scientists have long speculated that a causal relationship may exist between the almost synchronous development of the coal deposits late in the Carboniferous and extensive glaciation in other parts of the world.

The Carboniferous forests supported abundant animal life. The lakes and rivers were home to fish and shelly organisms. Amphibians splashed in the swamps and ambled on dry land. This was the time of the greatest diversification among amphibians, and for about 100 million years, they were the prominent animals on land. Included among them were large, armored animals, some fully terrestrial and others semi-aquatic; and limbless, snake-like forms; others had characteristics that foreshadowed the reptilian condition. Insects were everywhere, but particularly in the air. Modern insects are characterized by two very different types of wings: some insects, such as the dragonflies, have wings that are always held open; most, however, have wings that can be folded back and held close against the body when not in flight. These evolved in the Carboniferous. Nature, overcoming the constraints of land, had learned how to fly.

Quietly, simultaneously, a revolution took place. Amniotes, ancestral forms of all reptiles and mammals, emerged! Like the seed-bearing plants, amniotes probably evolved in areas with seasonal rainfall; their innovations, an evolutionary response to drought.

Amniotes evolved the key to vertebrate success on land: the amniotic egg, in which the developing embryo is surrounded by three membranes and a leathery or horny shell that prevent dehydration and allow for the exchange of air and waste gases. The latter function removed the size limitation that passive gas diffusion had imposed upon amphibian eggs. The reptilian egg could be considerably larger, the developing embryo could grow to a larger size before hatching, and the hatchling would not require an intermediate, aquatic growth phase before venturing onto land. An amniotic egg requires that fertilization take place internally before the membranes develop and the shell is laid down; water is no longer required as an intermediary agent. Desiccation, the problem that had forced amphibians to return to water to reproduce, no longer limited the distribution of vertebrates on land. So it happened, as the seas retreated from the continents and the climate around the world became increasingly dry, that reptiles became the dominant form of animal life on land.

The earliest known amniotes are preserved as the result of unique conditions that prevailed in the Late Carboniferous forests now exposed at Joggins, Nova Scotia.[9] Terrestrial animals lived in the lowlands that were dispersed among the coal swamps, perhaps seeking shelter in the hollow and decaying stumps of trees, perhaps becoming trapped and dying. Their remains were buried when their lowland homes were flooded; sediment settling out of the flood waters quickly buried and preserved the bones intact. The amniotes at Joggins were small and lizard-like with upright posture and well-developed jaw muscles for rapid jaw movement, anatomical characteristics that suggest they fed on insects. In fact, the evidence at Joggins suggests that the early success of amniotes may have depended upon the diversification and proliferation of insects.

The first amniotes were insect-eating carnivores, but specialized carnivorous and herbivorous groups emerged in the Permian. Two important groups can be recognized early in amniote evolution: the parareptiles, the ancestral reptiles among whom are found the ancestors of the dinosaurs; and the paramammals, the ancestors of the mammals.[10]

Early Permian vertebrate fossils document a complex terrestrial fauna comprising both amphibians and reptiles, the former at the peak of their evolutionary success, the latter rapidly evolving to fill new niches in the emerging arid terrestrial ecosystem. Most of the terrestrial amphibians may not have survived even into the Late Permian, but varied aquatic and semi-aquatic amphibians persisted throughout the Permian. The paramammals became numerous and diverse; included among them were many large and small land carnivores, semi-aquatic carnivores, and large herbivores.

It bears repeating that paramammals were the prominent amniotes of the Permian while the parareptiles, among them the typically small and bipedal carnivores, were not particularly significant or numerous by comparison. The emphasis is important because the fortunes of each group would be serendipitous in subsequent geological time.

The paramammals became increasingly mammal-like during Late Permian time, especially in the development of erect posture. Such

---

[9]The Joggins area is open to public visitation. See Jerry N. McDonald, 1992, *Old Bones and Serpent Stones—Volume 1: Eastern Sites.*

[10]I use the terms parareptile and paramammal (following Dale A. Russell, 1989, *An Odyssey in Time: The Dinosaurs of North America*) to identify the fundamental dichotomy of amniotes that has existed since their first appearance in the fossil record.

posture had dramatic evolutionary implications, for it brought about greater mobility. But it had a cost. It required both more energy and continuous energy to maintain the greater mobility. Producing energy also produced metabolic heat; ultimately, therefore, erect posture required that animals evolve a mechanism for controlling body heat. The fossil evidence for homeothermy is seen in the hard parts of the animals. In order to control its body heat, an animal had to be able to chew food into pieces small enough that it could be digested and turned into energy rapidly; the biochemical developments are recorded by the evolution of complex chewing teeth and elaborate jaw and muscle mechanisms.

The character of the terrestrial flora changed. The spore-bearing plants declined as the climate became drier, and the gymnosperms increased in number and diversity; the coal-swamp flora was being gradually transformed into the conifer-ginkgo-cycad flora characteristic of the Mesozoic.

The dramatic biological events of late Paleozoic time were set in a dynamic tectonic framework; the continents of the globe, having already been assembled into two giant continents, the northern Laurasia and the southern Gondwana, were drifting toward each other. When they collided, a single giant supercontinent, Pangea, was formed. The northern margin of Gondwana, present day Africa and South America, came into contact with the southeastern and southern margins of North America, extending the Hercynian Mountains southward and westward; the Ouachita Mountains are the remnants. The tectonic events of this time are visible in the rocks of Big Bend National Park in Texas [Site 29].

An analysis of late Paleozoic life on land emphasizes the impact of tectonics on evolution. From Late Carboniferous time onward, plants and animals became increasingly adapted to dry environments; an indication that as Pangea was being assembled, it became subaerially exposed and that large tracts of the interior of the supercontinent were desert. The greatest mass extinction of geological time marks the end of Permian time. It brought to an end the Paleozoic Era—the time of ancient life on earth. Mass extinction eliminated nearly all life in the oceans and drastically pruned animal life on land; the archaic land plants of the Paleozoic, already in decline, disappeared. Life on earth would be dramatically different from that time onward.

The decimation of life was overwhelming. Approximately 96 percent of all marine animals, both vertebrate and invertebrate, succumbed. The trilobites, far from numerous but still widespread since the Ordovician, became extinct; so did the tabulate and rugose corals. Shelly animals that lived in shallow marine environments—bryozoans, brachiopods, and crinoids—suffered the greatest losses. Nor were marine carnivores exempt: most ammonoids, five families of sharks, and eight families of bony fish disappeared; eurypterids, too, were formidable carnivores, but they were unable to survive. Many less well-known organisms also became extinct. The oceans of the world were left depauperate.

Mass extinction among land animals was less severe, but the impact was no less profound. Most of the fully-terrestrial amphibians disappeared, and the prominent amniotes, the paramammals, were greatly reduced. Extinction created many empty spaces in the terrestrial ecosystem into which the parareptiles would later evolve; by decimating the paramammals, extinction created an evolutionary opportunity for parareptiles and, ultimately, for dinosaurs.

What happened? Why? A number of events in the Permian are co-incident with mass extinction. During the late Paleozoic, tectonic processes had been assembling the landmasses of the earth to form large continents; by Late Permian time, all the landmasses had been joined to form the single gigantic continent, Pangea (Figure 11). Gondwana, the southern portion of Pangea, extended far into southern polar regions, and glaciers blanketed large portions of the southern hemisphere. A cause and effect relationship between the physical events and the biotic events seems the most obvious solution. There can be no doubt, for example, that a change in the climate to drier and cooler conditions had an impact on the biota, that the absolute amount of space for organisms in shallow-water environments in the oceans declined as continents coalesced and glaciers expanded. Yet, such historiographic explanations are simplistic and incomplete.

The tectonic events of the late Paleozoic took place over a protracted time, 40 or 50 million years. But the pattern of extinction does not parallel tectonic events, for mass extinction is only coincident with the final phases of mountain building. Superimposed upon the ebb and flow of extinction in the late Paleozoic, which continued into the Triassic, is a sudden and dramatic extinction, its spectacular nature best verified by how profoundly different life became afterwards. One may argue about speed and duration, about causes; but the fossils of animals that lived before and after the extinction document a profound event.

The Permian mass extinction is an important piece in the puzzle that emerges when global patterns of extinction are examined. For that reason, the role of catastrophic events, extra-terrestrial causes such as meteorite or comet impacts, to explain the mass extinction cannot be dismissed. It is especially true of the Permian mass extinction that paleontologists look at global patterns, because it is likely that direct evidence of events in the latest Permian may never be found. The latest Permian was a time of erosion world-wide, and the known sequences of Permian rocks are capped by an unconformity, a gap that reflects the erosion. The physical evidence of an extra-terrestrial impact, a sedimentary horizon rich in iridium for example, may not have been preserved; the sole evidence of such an impact may be the fact that the greatest mass extinction of organisms in all of geological time occurred then.

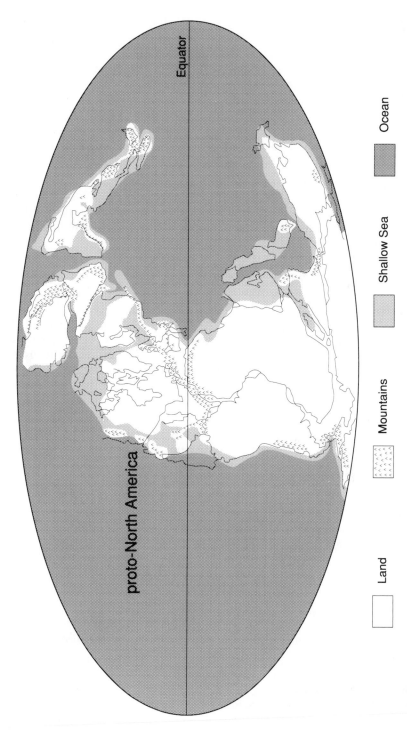

Figure 11. Paleogeography in Late Permian time. The continental landmasses of the earth had coalesced to form a single giant continent called Pangea.

Mighty Dinosaurs

Life was at its lowest ebb as the Mesozoic Era began, and the recovery was slow. Over time, from among the ruins of ancient life, evolution was able to fashion a splendid array of organisms unlike any that had gone before. Two major new groups of terrestrial vertebrates eventually emerged: the dinosaurs and the mammals. In the oceans from among the rare marine groups lucky enough to have survived, evolved new corals, new ammonoids, new plankton, and new fish. Only the gymnosperms among land plants remained unscathed by mass extinction, and their evolution proceeded in a more sedate fashion.

The Triassic Period was a time of evolutionary readjustment in the oceans. New species evolved sporadically, and eventually a distinctive invertebrate fauna emerged. Dominated by arthropods and mollusks, it had all the characteristics of a modern marine fauna.

A new phase of reef building was initiated when hexacorals (Scleractinia) made their first appearance in Triassic time; the corals are still extant, and reef building continues to the present time. The success of modern corals at reef building is based on symbiosis with algae, and only those corals that are symbiotic with algae produce reefs. The algae are photosynthesizers; they manufacture the food upon which the corals depend. The corals, in turn, by building a solid, three-dimensional structure, provide a home for the algae within the critical photic zone. The dominance of coral reefs was interrupted only once, in the Late Cretaceous, as nature experimented a fourth time with reef building. Some bivalve mollusks, the rudists, mimicked the corals and were able to build reefs, successfully displacing the corals from the warm, shallow, tropical waters and building all the reefs known from that time. Rudists became extinct at the end of the Cretaceous, however, and thereafter the corals resumed the reef-building role.

New marine predators emerged late in the Triassic. Large and rapid was the diversification of the ammonoids and the fish from among the survivors of the Permian mass extinction. The new ammonoids were the ammonites, shelled relatives of modern squid but superficially resembling the modern *Nautilus*, and they became the dominant invertebrate predators. Some grew to be gigantic in size, developing coiled shells that grew up to two meters in diameter. The large *Placenticeras* at the Kremmling Cretaceous Ammonite Locality in Colorado [Site 20] belongs to this group. Fish reemerged as the most numerous and varied of the marine vertebrates, but their significance is overshadowed by the motley crew of reptiles that returned to the oceans.

Six groups of reptiles made the transition to a marine existence, each evolving unique ways of living in the oceans. Ichthyosaurs, plesiosaurs, mesosaurs, and mosasaurs became totally aquatic; the only constraint from their terrestrial heritage was that of obtaining oxygen from the air rather than the water. Each adopted different anatomical changes to facilitate swimming. While plesiosaurs, for example, modified their limbs into great paddles and moved through the water by using a rowing motion, ichthyosaurs developed streamlined, tuna-shaped bodies and were fast and agile swimmers. The skeletons at Berlin-Ichthyosaur State Park in Nevada [Site 41] confirm that ichthyosaurs were unable to survive when removed from the water. Only crocodiles and turtles did not become totally aquatic; they, alone among the aquatic reptiles, survive to the present.

When life on land began to recover in the Triassic, the pattern of prominence that had characterized the Permian was reversed. The paramammals had been decimated; only three groups survived, and they were slow to recover. But, from among the surviving paramammal carnivores, the first mammals appeared. They remained small in size throughout the Mesozoic and were but a minor component of the terrestrial vertebrate fauna. The vigor that paramammals had exhibited in the Permian accrued, in the Triassic, to the parareptiles; the lizard- and crocodile-like reptiles became the dominant land animals. The remains of phytosaurs, aquatic crocodile-like carnivores, are common in western North America, especially in the Late Triassic Chinle Formation which is exposed at Petrified Forest National Park in Arizona [Site 32].

The mighty dinosaurs had humble beginnings, evolving from parareptilian ancestors in the Early Triassic. The first dinosaurs were small carnivores—bipedal, swift and agile. They are readily recognized as fossils by their special adaptations to bipedal locomotion: a rigid backbone, alignment of all the bones of the hind limb for fore-and-aft motion, specialized muscle attachments to increase the strength and efficiency of muscles. Accompanying these changes was the characteristic loss of bones of the fourth and fifth digits of both fore and hind limbs. One of the oldest known dinosaurs is *Coelophysis,* the best record of which is from the *Coelophysis* Quarry at Ghost Ranch Conference Center in New Mexico [Site 31]. The quarry is in the Chinle Formation, and the first dinosaurs are, therefore, contemporaneous with fossils from Petrified Forest National Park and with the first mammals.

Mass extinction, this time in the latest Triassic, once again rearranged the equilibrium among living organisms. Although many marine invertebrates became extinct, the terrestrial extinctions are much more compelling, because this time, the ruling parareptiles were decimated. They were the dominant land vertebrates of the time; only when they were eliminated did the dinosaurs display their full potential.

The informal term dinosaur refers to two great groups of reptiles, the Order Saurischia and the Order Ornithischia. They are distinguished by the structure of their hip bones: the Saurischia had hip bones characteristic of reptiles; the Ornithischia had hip bones resembling those of birds. The names refer to the different morphologies of the ischium, a hip bone; *saur* comes from the Greek word, *sauros*, for lizard; *ornithos* from *ornis*, the Greek word for bird.

The reptile-hipped dinosaurs were the first dinosaurs to appear; and late in the Triassic, two very different groups of animals, Suborder Theropoda and Suborder Sauropoda, could be recognized among them.[11] Seventeen families are included in the Theropoda, the only dinosaurs to retain the carnivorous lifestyle of their ancestors. Some species were small and bird-like;[12] others, such as *Tyrannosaurus*, became giants; but all were bipedal and agile. The Sauropoda contrasted sharply, for they were quadrupedal herbivores! Familiar forms include the brontosaurs, some of the largest animals ever known on earth. Sauropod giants are best known from the Late Jurassic, in rocks of the Morrison Formation which are exposed over a wide area in Utah, Colorado, and Wyoming on the eastern flanks of the Rocky Mountains. Dinosaur National Monument in Utah [Site 40] features the single largest exhibit of *in situ* sauropod bones.

The bird-hipped dinosaurs, more complex than the reptile-hipped dinosaurs, are subdivided into four suborders. The first to appear in the fossil record, late in the Triassic, were the Ornithopoda, which evolved to be the dominant bird-hipped dinosaurs of the Jurassic and Cretaceous. All the ornithopod dinosaurs were bipedal, but unlike the bipedal theropod dinosaurs, they were vegetarian. The best known ornithopods are the duck-billed dinosaurs, the Family Hadrosauridae, found in many rocks of Late Cretaceous age. Some localities, such as the Willow Creek Anticline sites in Montana [Site 12], even preserve hadrosaur nesting sites. Three other kinds of bird-hipped dinosaurs evolved during the Mesozoic, each differing from the ornithopods in that they reverted to a quadrupedal stance. Nonetheless, even a cursory look at their skeletons, which bear markedly shorter fore limbs than hind limbs, reveals their bipedal ancestry. The Stegosauria, the plated dinosaurs, evolved in the Jurassic; the armored dinosaurs, the Ankylosauria, in the Cretaceous. The last to appear, late in the Cretaceous, were the Ceratopsia, the great horned dinosaurs exemplified by *Triceratops.*

---

[11]The classification of dinosaurs is complex. In order to accurately represent the similarities and differences among them, an additional level of classification, intermediate between order and family, is used. The Order Saurischia and the Order Ornithischia have been subdivided into suborders into which closely related families are placed.

[12]The analogy, like the relationships, can be confusing. Many of the theropod dinosaurs, although reptile-hipped, were very bird-like in morphology and appearance. In contrast, the bird-hipped dinosaurs had little resemblance to birds, yet the name of the bipedal group (Ornithopoda) is based on the bipedal or bird-like posture. To complicate the analogy further, the evolution of birds themselves can be traced back to theropod (and therefore, reptile-hipped) dinosaurs or to a common ancestry with theropod dinosaurs.

Dinosaurs, the pre-eminently successful land vertebrates, were remarkable in every aspect of their evolutionary history. They included carnivores and herbivores, large animals and small. Many were swift and agile, many lived and foraged in herds; some evolved the capacity to regulate their body temperature, others protected their nests and cared for young hatchlings. They were exceedingly diverse, the most sophisticated among the reptiles.

Reptile success extended into the air. Pterosaurs, the flying reptiles, appeared with the dinosaurs; and later in the Triassic, the first birds evolved from theropod dinosaurs.[13] For about 180 million years reptiles were the dominant animals on earth, occupying habitats on land, in air, and in water.

Reptiles were dominant, but they were not alone. Mammals evolved and lived side by side with dinosaurs from the first appearance of each in the Late Triassic; their evolution and diversification during the Mesozoic Era was comparable to that of the dinosaurs, but unlike the dinosaurs, they remained small and inconspicuous. Lilliputians in a world of giants, mammalian fossils were overlooked for years by paleontologists preoccupied by the abundance of large dinosaur bones.

Among the characteristics that distinguish modern mammals from reptiles is their reproductive strategy, the bearing—and nursing—of live young. Modern mammals, however, comprise two distinct groups, each with a distinct mechanism for bearing their young. Marsupial mammals are born at an immature stage and are then nurtured by the mother in a pouch (marsupium) for an extended period of time. The placental mammals developed an internal structure, the placenta, by means of which the young are nurtured for an extended time *in utero*. Vestiges of the ancestry of mammals are seen in the living monotremes, the platypus and echidna of Australia, which are egg-laying mammals.

The long story of the evolution of mammals during the Mesozoic Era is written in their teeth. Mammalian teeth, like those of reptiles, are constructed of durable material; they are the most common remains of mammals in the fossil record. In contrast to the simple, basic reptilian tooth pattern, which consisted of conical teeth distributed along the jaws with new teeth replacing worn ones in a regular pattern continuously throughout life, the basic mammalian pattern that evolved was complex. Teeth of different shapes to serve different functions were positioned along the jaws and interlocked in a specific way. The teeth had an interdependent function; they were replaced only once, the second set designed to serve the adult animal for life. The pattern was already present in the Late Triassic mammals; the subsequent modification in the morphology

---

[13]The origin of birds is not totally resolved. Recent evidence suggests that birds may be as old as dinosaurs and may be derived from an ancestral form that is common to both birds and dinosaurs.

of teeth documents the evolution of complex physiology among the mammals.

It must be stressed that many of the characteristics that define mammals—live birth, high metabolic rate, internal temperature regulation—are not subject to fossilization. That mammals attained these various levels of development is inferred from changes in tooth morphology over time. For example, because the fundamental tooth structures that distinguish marsupial and placental mammals were already present in Early Cretaceous time, it is likely that physiological distinctions were also present. On the other hand, the presence of teeth of placental mammals in Early Cretaceous time does not prove that the placental level of reproduction had also been achieved.

The characteristic plants of the Mesozoic Era were gymnosperms, the familiar conifers in modern environments. Conifers and the extant but less familiar cycads and ginkgoes formed extensive forests during the Mesozoic Era. Ferns were plentiful. In fact, the Mesozoic plant world would have appeared odd to us, not because the plants were unfamiliar, but because certain modern plants had not yet evolved. In the Early Cretaceous, however, flowering plants with covered seeds, the angiosperms, first appeared, and the structure of plant communities gradually changed.

Mass extinction 65 million years ago marks the end of the Mesozoic Era. It eliminated many groups of animals, most visibly the dinosaurs; like the Permian mass extinction, it profoundly altered life on earth. Land animals were the most severely affected: dinosaurs and flying reptiles disappeared; so did many mammals, particularly the marsupials on the northern continents. Many plants disappeared; the composition of the land flora changed from dominantly gymnospermous to dominantly angiospermous.

Marine life was less severely devastated by mass extinction, but many prominent groups of organisms were culled. Marine reptiles, except for crocodiles and turtles, became extinct. The ammonoids, dominant invertebrate carnivores with a long evolutionary history, had been particularly successful at retaining survivors during periods of severe devastation, including the Permian mass extinction. At the end of the Cretaceous, however, they were eliminated for all time. Among the large invertebrates and microscopic marine organisms alike are many that did not survive.

The extinction of the dinosaurs has captured public imagination and scientific curiosity. Intense research over the last decade has been motivated by the search for a cause; the data are diverse: a gradual decline in the diversity of dinosaurs over time, high concentrations of iridium in key marker beds, charcoal in sedimentary rocks that could only have come from massive fires. Various factors have been invoked to explain the death of dinosaurs, but be they environmental (such as a world-wide drop in temperature), tectonic (such as mountain building on the west coast of North America), or extra-terrestrial (such as asteroid impacts), no consensus has been reached.[14]

The Cretaceous mass extinction is an event of larger proportion than the extinction of the dinosaurs, larger because many other organisms were also extirpated and larger because evidence of extra-terrestrial causes is strongest. Understanding the Cretaceous mass extinction continues to provide insight into global patterns of evolution and extinction.

Extinction at the end of the Cretaceous totally restructured the biosphere, bringing to an end the time of middle life. As at the end of the Permian, life would never be the same again.

Mammals in a Modern World

Selected mammals survived the Cretaceous decimation, but the new world order around them was both strangely familiar and totally foreign. The plants were familiar, for the change in the flora that began in Early Cretaceous time continued unscathed into the Tertiary. The insects of the Paleocene were like those of the Late Cretaceous. Most of the vertebrates, however, were gone.

The early Cenozoic world inherited by the mammals was beginning to look modern. During the dynasty of the reptiles, Pangea had slowly begun to break apart, and the continents of the northern hemisphere had separated from the continents of the south. The Atlantic Ocean, already present in the southern hemisphere, was only beginning to open between the continents of the north; North America and Eurasia were to emerge as distinct landmasses. South America and Africa were now isolated continents moving away from each other, separated from an Australia-Antarctica landmass and from the Indian subcontinent. India and Africa were moving northward on a collision course with Eurasia.

The fragmented continents inherited distinct suites of mammals. Only placental mammals survived into the Cenozoic Era on the large North American-Eurasian landmass. Australia-Antarctica had no placentals, only the marsupials and primitive egg-laying mammals. South America and Africa each had a unique assortment of founding mammals that remained distinct only until land connections were reestablished with continents of the northern hemisphere. The faunas of Europe, Africa, and Asia commingled; mammals migrated into and out of North

---

[14]The reader interested in pursuing the topic of Cretaceous extinction is referred to the bibliography in Section III. Authors of the various books on dinosaurs address extinction from their own perspectives (for example, Dale Russell and Robert Bakker). Other authors address the issue of extinction directly (for example, David Raup).

America across Panama and across the Bering region. Mammalian evolution during the Cenozoic Era was like an intricate experiment; the variable factors over time, isolation and interchange. The history of mammals on each continent, therefore, is unique.

Tectonic events that began early in the Cenozoic Era initiated global climatic change. The continents drifted toward the poles, and circumpolar ocean currents developed. The climate gradually shifted from the equable, warm to tropical conditions prevalent at the beginning of the Cenozoic to unstable conditions characterized by extensive glaciation at its close. Three times during the first 25 million years or so of Cenozoic time, annual mean temperatures plummeted, at least in North America, each time by about 10 degrees Celsius, and then climbed again to previous mean temperatures. A fourth drop in the annual mean temperature early in Oligocene time was precipitous. That time temperatures did not recover. Glaciation of Antarctica began in the Miocene, and during Pliocene time, glaciers began to developed in North America.

Floras of the Cenozoic evolved rapidly in response to changing environmental conditions. Forests of angiosperms and gymnosperms, such as those preserved at Yellowstone National Park in Wyoming [Site 13], dominated the landscape during Paleocene and Eocene time; they differed from forests of the Cretaceous only in the relative proportions of constituent plants. Flowering plants, which had emerged midway through the Cretaceous, had not been affected by the mass extinction that had decimated so many organisms; instead, they came to dominate the flora at the expense of gymnosperms. Most notably, during Eocene time, specialized flowering plants, the grasses, were already present. They remained insignificant among the diverse plants that constituted widespread forests, a minor component of the flora, until late Oligocene and Miocene time. Then, as temperatures dropped and precipitation declined, they flourished. Grasslands replaced the forests, and grasses became the dominant plants of temperate and dry regions.

The success of vertebrates throughout their evolutionary history is based in their persistence as predators. The first mammals, like the first reptiles and the first amphibians before them, were predators. From a predatory insectivorous founding stock, mammals rapidly evolved a variety of herbivorous and carnivorous species that came to occupy a wide variety of ecological niches.

The pattern of rapid mammalian evolution early in the Tertiary is broadly parallel to the pattern of reptilian evolution approximately 180 million years earlier. As did the reptiles before them, mammals diversified widely on land, conquered the oceans of the world, and took to the skies. The radiation occurred rapidly in the early Cenozoic; within the first 10 million years, most modern subdivisions (Orders) of mammals including carnivores, whales, bats, odd-toed ungulates (Order Perissodactyla, which includes the modern horses and tapirs) and even-toed ungulates (Order Artiodactyla, cloven-hoofed animals such as bison and deer), elephants, and rodents appeared. In the Oligocene, there was a trend toward gigantism among the mammals, evident from the many species of large Oligocene mammals that are known. The diversity of Oligocene mammals preserved in the White River Formation at Badlands National Park in South Dakota [Site 17] includes perissodactyls such as the titanothere, *Brontops* and the horse, *Mesohippus*; and artiodactyls such as *Merycoidodon* and *Leptomeryx*.

The widespread development of grasslands beginning in the Miocene provided the impetus for the evolution of highly specialized grazing herbivores. The fossils at Agate Fossil Beds National Monument in Nebraska [Site 19] are characteristic of that time. Grasses, high in cellulose which is tough to chew and difficult to digest, are a low-quality source of food. In spite of that, odd-toed ungulates and even-toed ungulates evolved mechanisms for consuming and processing grasses as their primary source of food. Each evolved specialized teeth and sophisticated mechanisms for digestion that allowed them to extract maximum nutrition from their poor forage.

Horses are the most specialized grazers among the odd-toed ungulates. Their teeth evolved into massive grinders made up of complex infolding of the internal structures of the teeth. Because the teeth were elongated and evergrowing, they would last a life-time. With such teeth, horses could consume large quantities of highly abrasive food. The stomach became enlarged and the digestive tract greatly lengthened. A specialized fermentation chamber evolved in the intestine to house cellulose-digesting bacteria, which by breaking down the cellulose, enabled horses to extract a higher proportion of nutrients from their food.

Among even-toed ungulates, antelope and bison are efficient grazers. Like horses, they have evolved elongate grinding teeth and an elongate digestive tract. In contrast to horses, however, they have a ruminant stomach, one chamber of which functions as a huge fermentation vat housing cellulose-digesting bacteria. The ruminant stomach is more efficient than the fermentation chamber in horses and is capable of extracting maximum nutrients from cellulose-rich grasses. The increased efficiency means that the even-toed ungulate grazers require proportionately less food than do horses and that they can survive on forage of lower quality than can horses.

Rodent herbivores evolved in the Tertiary. Small and seemingly invisible, rodent herbivores are not to be dismissed, for they were and are significant components of terrestrial ecosystems. Rodents, in total, are greater consumers of plant material than are all of the ungulates combined. Furthermore, they support a complex community of predators that includes birds and reptiles as well as mammals.

Carnivores are the ultimate consumers in nature, but there are fewer carnivores in any ecosystem than there are herbivores, and fewer kinds

59

of carnivores. The number of individuals within a species that an eco-system can support decreases the higher the position of that species on the food pyramid; thus, for example, it takes many rabbits to maintain a single owl during its lifetime. Among the land mammals in Tertiary time, however, the diversity of carnivores was higher than at any other time among any group of vertebrates. Some were more specialized than others, and many species are now extinct, but most are identifiable within well-known modern families: Canidae, the dogs; Felidae, the cats; Mustelidae, the weasels; Ursidae, the bears.

Never before had carnivores evolved such highly specialized mechanisms for hunting, stalking, and killing prey. The saber-tooth cat *(Smilodon)* is well known, the epitome of carnivore specialization. Its blade-like canine teeth were deadly, its cheek teeth modified to sharp shears that readily sliced meat into bite-sized pieces. Individuals of *Smilodon* were among the many carnivores attracted to the tar pits at Rancho La Brea in California [Site 43] where their skeletons are preserved in large numbers.

Glaciation introduced the modern world. Glaciers began accumulating in polar regions and in the mountains as early as five million years ago; then they spread southward across the continent, eventually covering as much as two-thirds of North America. Massive glacier ice advanced numerous times, each advance punctuated by smaller-scale retreats of ice and separated from the next advance by an interglacial interval—a time when the ice melted, and warm and equable climates prevailed. We live in an interglacial time, but this one is different than previous ones for the climate is more distinctly harsh and strongly zoned.

Between 15,000 and 11,000 years ago, human beings first colonized North America. Following some two million years of evolution in Africa and then in Europe and Asia, humans developed the cultural skills necessary to survive in harsh, northern climates. One of the northern areas they inhabited was Beringia, a refuge comprising the Bering Strait and land from the two continents on either side that was subaerially exposed when the formation of glaciers lowered sea level. Beringia—a half-way land leading to the heart of North America. Migration to North America was unlikely to have been a conscious decision, but an event contingent upon climatic factors and the presence of game animals.

Almost synchronously with the dispersal of humans, many animals in North America, predominantly the large mammals, became extinct. A coincidence? Some paleontologists and archeologists think not and argue that the first migrant humans hunted the large mammals to extinction. Hunt the large mammals they did, as spear points embedded in fossil bones attest; but hunt them to extinction? We may never know. In the meantime, early human habitation and hunting sites, such as Black-water Draw in New Mexico [Site 26] and Lubbock Lake in Texas [Site 27], are becoming better known, and our understanding of the big-game hunters and their world becomes clearer.

The Making of North America

A physiographic map of North America clearly shows that the continent is made up of distinct physiographic units, each uniquely characterized by features of topography and geology (Figure 12). The overall structure is deceptively simple: a central core is blanketed by layered rocks and encircled by mountains. The core of North America is the Precambrian Shield, exposed in Canada in a great arc around Hudson Bay. It extends southward beneath the layers of sedimentary rocks that make up the interior platform of the continent. The Shield is flanked by great linear belts of mountains: the Appalachian Mountain Belt on the east and southeast margin, the Cordilleran Mountain Belt on the western margin. The Innuitian Mountains in the Arctic are an extension of the Cordillera to the northeast and into Greenland; the Appalachian Mountains once extended far to the southwest, but only remnants remain visible. The Coastal Plain is defined by the recent sediments being deposited on the shallow eastern and southern margins of the continent. The geological events in each unit constitute discrete chapters in the history of North America.

The Precambrian Shield, where some of the oldest known rocks in the world are found, is the oldest portion of North America. Not a simple homogeneous unit, it comprises four main structural provinces; the oldest, the nucleus around which the continent grew. Each of the other structural provinces formed when a moving plate of earth crust collided with the nucleus; each collision extended over a very long period of time and was characterized by metamorphism, deformation of rocks, and massive mountain building. Approximately three billion years of earth history passed before the core of North America, a landmass recognizable as proto-North America, was assembled.

The rocks of the Precambrian Shield record the first 85 percent of the history of North America, but that history is very difficult to elucidate. The mountains that once existed have been eroded away; the rocks that remain have been altered by deformation and metamorphism to such an extent that the original form of many is obscured. Organic remains are rare and difficult to interpret. In sharp contrast, the thick, layered rocks that formed over the last 600 million years are exceedingly well preserved and widely distributed. Many are unaltered, and many contain abundant fossils. The comparative abundance of clues allows geologists and paleontologists to document a rich and complex history of more recent geological time.

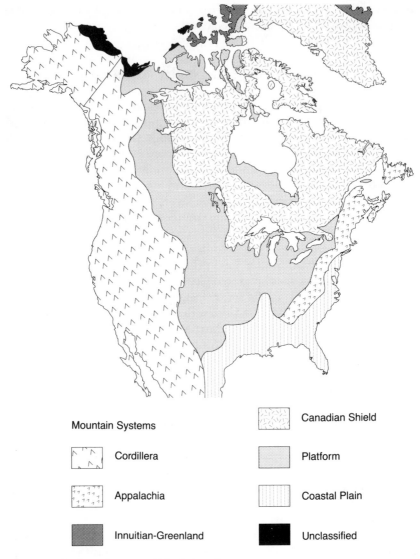

Mountain Systems

Cordillera

Appalachia

Innuitian-Greenland

Canadian Shield

Platform

Coastal Plain

Unclassified

Figure 12. Physiographic map of North America.

The Precambrian Shield, the bedrock of North America, can be viewed as the foundation upon which the layered rocks present today were deposited. Six times over the last 600 million years, at irregular intervals, ocean waters flooded and then receded over the landmass that was proto-North America. Each time, the advancing epicontinental seas deposited layers of fossil-rich sediment on top of pre-existing rocks. Great expanses of limestone rock are silent testimony to the ancient seas.

The fossils at Stonewall Quarry Park in Manitoba [Site 1] document one such ancient sea; those at Minnetonka Cave in Idaho [Site 14], another. When the seas receded, terrestrial organisms spread out to inhabit the land, and wind and water began inexorably to erode the sediments. In some places, rivers redeposited the sediment in deltas and swamps; in others, the wind deposited massive sand dunes. In Grand Canyon National Park in Arizona [Site 33], marine and terrestrial deposits are preserved in vertical sequence. Fossiliferous, layered rocks once blanketed most of the core of North America, but agents of erosion—wind, water, and ice—have removed many of them. Those that remain cover the southern and western Shield.

The rocks of the Appalachians record, over a time of about 400 million years, a series of orogenies or collisions of landmasses that built great mountain chains and ultimately enlarged the proto-North American continent. Fragments of earth crust that today constitute North America and Europe moved together and collided, eventually forming a single, large continent. In the process, a large intervening ocean, the Iapetus Ocean, was swallowed up,[15] and the massive Hercynian Mountain Chain, extending from Scotland in the north, southward through Newfoundland and into the Appalachians of the United States, was formed.[16] The giant northern continent, Laurasia, had been assembled.

By Late Carboniferous time, all the landmasses of the earth had been assembled to form two giant continents, the northern Laurasia and the southern Gondwana. The continents were moving toward one another during Late Carboniferous and Permian times, and a portion of land that was ancient Africa collided with the part of Laurasia that is now southeastern North America. Northern Gondwana, land that is now Africa and South America, collided with Laurasia and extended the Hercynian Mountains southward and westward. The Ouachita Mountains of Arkansas, Oklahoma, and central Texas are the remnants of the western extent of the Hercynian Mountains. At the end of Permian time, some 250 million years ago, the collision complete, all of the continents of the earth were assembled into one gigantic landmass called Pangea.

The western margin of North America was relatively featureless for most of Paleozoic time while mountains were being built along the east coast. The Cordillera had not yet formed, and the continental landmass gently yielded westward to continental shelf and open oceans. The interior of the continent was periodically flooded, for any rise in sea level was unimpeded. The mountain belt that defines the modern size and

---

[15]The Iapetus Ocean was the forerunner of the modern Atlantic Ocean. It is named after Iapetus, the father of Atlas (from Greek mythology), for whom the Atlantic Ocean is named.

[16]For a more complete history of the geology of eastern North America see Jerry N. McDonald, 1992, *Old Bones and Serpent Stones—Volume 1: Eastern Sites.*

shape of western North America is the product of three major orogenies separated by periods of quiescence, each mountain-building event comprising several distinct episodes. The effects of collision are manifest in rocks over a wide geographical area, and in many areas, the local geological events bear a superimposed regional imprint that derives from mountain building to the west.

The oldest mountain-building event on the western margin of North America is the Antler Orogeny, which took place in two phases during Paleozoic time. About 350 million years ago, in Late Devonian and Early Carboniferous time, long before Pangea was assembled, tectonic processes became active along the portion of western North America that is now west-central Nevada. A chain of volcanoes, which had been located offshore in the ocean, collided with the margin and produced highlands called the Antler Mountains. In Late Permian time, some 100 million years later, the volcanoes themselves became sutured onto the continent, thereby increasing the size of North America. The rocks that were emplaced at that time form the Klamath Mountains and the Sierra Nevada in California.

The Four Corners area of the United States (Utah, Colorado, Arizona, and New Mexico) has had a particularly interesting geological history. The region records two distinct, superimposed phases of mountain building: one developed in the Late Paleozoic as Pangea was assembled; the second, much later in the Early Tertiary, as the Cordillera was built. During the first phase, the area was affected by Hercynian events when the Ouachita Mountains were formed. The tectonic activity produced a series of highlands and basins aligned approximately northwest to southeast, the highlands called the Ancestral Rocky Mountains. The terminology is somewhat confusing since the highlands were formed by the building of the Ouachita Mountains, not by the building of the Rocky Mountains, in an area that was subsequently altered by events that built the Rocky Mountains.

No sooner had Pangea been assembled than it began to break apart, rifting slowly at first but with increasing speed as time went on. Rifting began in the Triassic, and the modern North Atlantic ocean began to form. North America began to drift northwestward. It collided along its western margin with adjacent plates that made up the Pacific Ocean; a subduction zone, which extended south to include the western margin of South America, developed along the western margin of the continent. Mountain building began along the entire front.

Geological activity along the western margin of North America has been essentially continuous since Jurassic time, marked by alternating periods of intense tectonism and relative quiescence. Nonetheless, mountain building has not been continuous along the entire length of the mountain belt; different centers have been active over different periods of time.

The area that is now central Nevada was the focus of intense mountain building beginning about the middle of Jurassic time. Oceanic crust slipped beneath the continental landmass, and rocks were compressed, crumpled, and uplifted to form mountains. This tectonic activity is referred to collectively as the Nevadan Orogeny.

Simultaneously, ocean waters began flooding onto the land and eventually formed a large epicontinental sea, the Sundance Sea; which at its maximum, extended from central Arizona and New Mexico northward into southernmost Alberta and Saskatchewan, from Nevada eastward to include most of North and South Dakota (Figure 13). Along its western shoreline were the mountains of the Nevadan Orogeny. An

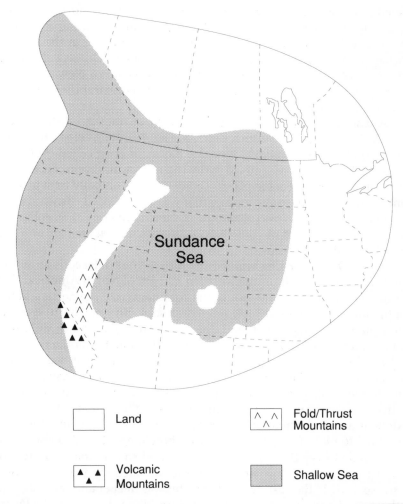

Figure 13. The Sundance Sea. At its maximum extent during the Jurassic, the Sundance Sea flooded a vast region in the interior of North America.

extraordinary volume of sediment was eroded from the mountains as they formed, then carried by rivers to and deposited in the Sundance Sea. Eventually, in the Late Jurassic, the volume of sediment was so large that the Sundance Sea was filled in, and the area was crossed by streams and dotted with lakes. The deposits that finally filled the Sundance Sea are today called the Morrison Formation, rocks that extend over an area of about one million square kilometers and are famous for their abundance of dinosaur bones. The great dinosaur rush of the late nineteenth century was concentrated on the Morrison Formation; among the sites excavated then that are still producing dinosaur bones is the Garden Park Fossil Area near Cañon City in Colorado [Site 23].

Mountain building in the Central Rockies of the United States continued into and throughout the Cretaceous, expanding northward into Idaho and Washington and eastward into Wyoming to encompass a very broad area. The Cretaceous phase of mountain building in the Central Rockies is often given a separate name, the Sevier Orogeny. Volcanic activity was intense, especially in Idaho and northern Nevada. Rocks were uplifted, folded, and thrust eastward across central Utah, western Wyoming, and west-central Montana.

A distinctive style of mountain building characterized the Cordilleran Mountain Belt in western Canada. The mountains along the coast comprise at least two unique, elongate blocks of crust called exotic terranes. Each block was assembled in the Pacific Ocean: slivers of continental crust, carried on Pacific plates, collided and eventually built a small continental mass, an exotic terrane that was then carried eastward until it collided with the western margin of North America. The collisions sutured the exotic terranes onto North America, one in the Late Jurassic, the other in the Cretaceous.

Sea levels were rising world-wide during the Cretaceous as mountains were being built on the western margin of North America. The oceans flooded into the interior of the continent from the north and the south, forming extensive epicontinental seas. The northern sea was dominated by deposition of mud, the southern by limestones and reefs. In Late Cretaceous time, sea levels were high enough that the northern and southern seas were joined, and the Western Interior Seaway was formed. Great thicknesses of sediment, derived from the rising mountains to the west, accumulated in the sea. The western margin of the sea shifted back and forth over time in response to the variable volumes of sediment and to fluctuations in world-wide sea level: sediment tended over time to fill in the sea and shift the shoreline eastward, a rise in sea level flooded over the sediments and shifted the shoreline westward. Deposits of rivers, deltas, and swamps dominated the western margin of the seaway. To the east, where the water was deeper and the sediment influx less, limestones were deposited. Some deposits, such as the Pierre Shale at the Kremmling Cretaceous Ammonite Locality in Colorado

Figure 14. Life and death in the Western Interior Seaway. Plesiosaurs and ammonites were common inhabitants of the Western Interior Seaway in Late Cretaceous time. Many of the remains that once littered beaches, such as those illustrated in the diorama at the Saskatchewan Museum of Natural History, are now preserved as fossils.

[Site 20], contain evidence of rich and abundant marine life; others, such as the Judith River Formation of Dinosaur Provincial Park in Alberta [Site 2], yield an unsurpassed array of dinosaurs. Late in the Cretaceous, renewed mountain building increased the supply of sediment to the interior, and sea levels world-wide were falling; the seas withdrew completely from the western interior of North America and have not returned. The marine organisms, like the dinosaurs on land, were culled by extinction (Figure 14).

Renewed tectonics, the Laramide Orogeny, began late in the Cretaceous and persisted throughout the Tertiary. Over most of the Cordilleran Mountain Belt, particularly in the Canadian Rockies in the north and the Sierra Madre of Mexico in the south, mountain building consisted of volcanic activity, and folding and eastward thrusting of rocks. The geographical extent and duration of the volcanism is documented by numerous fossil localities: John Day Fossil Beds National Monument in Oregon [Site 7] is Eocene to Miocene in age, Florissant Fossil Beds National Monument in Colorado [Site 22] is Oligocene, Ginkgo Petrified Forest State Park in Washington [Site 6] and Stewart Valley Paleontological Area in Nevada [Site 42] are Miocene, Hagerman Fossil Beds National Monument in Idaho [Site 9] is Pliocene.

The central portion of the Cordillera, extending from Montana to Arizona but focused on Colorado, experienced a different style of tectonism. Here, huge blocks of basement rock were uplifted, and the Ancestral Rocky Mountains were elevated again. A series of highlands and basins were formed; the latter were sites of accumulation of great thicknesses of sediment, now some of the richest fossil beds of Cenozoic time. The Green River Basin, for example, accumulated sediment throughout Eocene time, and the Green River Formation is famous world-wide for its exquisitely preserved fish, birds, and bats. The sequence of sediments is exposed at Fossil Butte National Monument in Wyoming [Site 15].

Tectonic events determined, in one way or another, both the nature of the fossil localities and their distribution. Some fossils, such as the petrified trees in Petrified Forest National Park in Arizona [Site 32], are preserved because volcanoes distributed ash on the surrounding countryside, ash rich in silica that dissolved readily in ground water and petrified bone and wood. Other fossils, such as the soft-bodied organisms in the Burgess Shale of British Columbia [Site 4], are exposed to our view because mountain building elevated the fossiliferous rocks high above sea level. The overprint of tectonics in the western Cordillera can be seen in all fossil sites of relevant age in western North America.

Comparison of dinosaur bonebeds illustrates the impact of tectonics on fossilization and underscores the inter-relationship that has existed through time between the physical earth and the biological world. Late Jurassic dinosaur bonebeds in the Morrison Formation in the western United States have at least one aspect in common with Late Cretaceous dinosaur bonebeds in the Judith River Formation in western Canada. In each case, the bones were buried rapidly and in large numbers because the mountains that were built in the Cordillera (the Nevadan Orogeny in Jurassic time; the Laramide Orogeny in Late Cretaceous time) shed much sediment, sediment that was carried eastward by rivers flowing into an epicontinental sea. In each case, the fossil-bearing strata were buried by younger deposits. Mountain building subsequently uplifted and exposed the Morrison Formation, as at Dinosaur National Monument in Utah [Site 40]; but the Judith River Formation is exposed because erosion has carved down through the overlying sediments and formed the badlands at Dinosaur Provincial Park in Alberta [Site 2].

Tectonic activity on the western margin of North America has not ceased. Volcanoes, such as Mount St. Helens, continue to erupt. Earthquakes are a constant threat. Geysers boil at Yellowstone National Park. Hot springs are a surprising discovery on many isolated mountain hikes. The beauty of a magnificent landscape belies the power hidden within.

Section II

Fossil Localities Interpreted for the Public

Fossil Localities
Interpreted for the Public

Section II documents 45 fossil localities in western North America that are open to the public. The sites are located west of 100 degrees West Longitude; the Province of Manitoba and the states to the south that are bisected by that line are considered western. Although the sites are widely distributed across western Canada and United States, they are presented on the following pages in approximate order from north to south. The number of each site in the text corresponds to the site number on the map that follows (Figure 15).

The documentation for each site provides two distinct kinds of information. The interpretive information describes the geological setting and paleontological significance of the site and its fossils. The information on public access and use includes directions to the site and, where appropriate, to specific fossil exhibits on the site; the recreational and educational facilities at the site are described, and sources of additional information are listed. Location maps are provided for all the sites. All 45 sites are managed in a way that encourages public interest in fossils. Twenty-eight are located within parks, either national, state/provincial, or regional/local. The management and development of the fossil resources have been incorporated within the mandate of parks whose twin objectives are protection and public enjoyment. The same rules apply to fossils as to other natural aspects of the parks; specifically, collecting of any materials is prohibited. Ten sites in the United States are located on public lands managed by the Bureau of Land Management. These, given the different mandate of the Bureau of Land Management compared to that of parks, tend not to be structured as strongly or interpreted in as much detail as are fossil sites in parks. One site is in a national forest. Six are privately owned either by individuals or private organizations; visitors are encouraged, but permission is required, and as a rule, visitors will be accompanied to the fossil site. *Note that collecting fossils at any of these sites is strictly prohibited.*

The sites in western North America feature a wide variety of fossils and span approximately 1.0 billion years of earth history. They represent

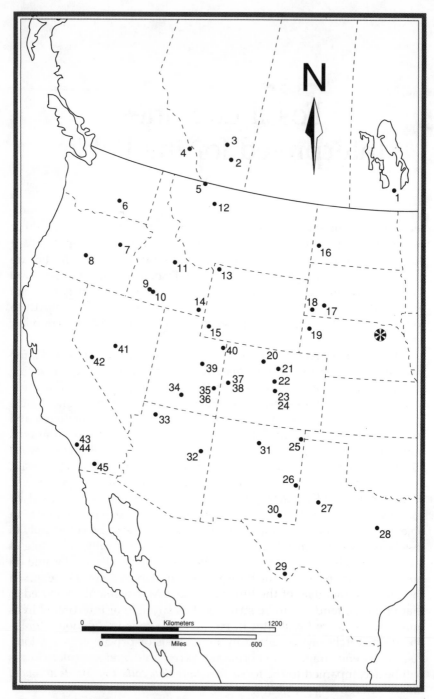

Figure 15. Map of western Canada and United States showing location of interpreted fossil localities. The sites are identified on the facing page. Numbers on the map correspond to those of the sites documented in Section II.

Interpreted Fossil Localities
in Western Canada and United States

1. Stonewall Quarry Park, Stonewall, Manitoba
2. Dinosaur Provincial Park, Brooks, Alberta
3. Dinosaur Trail/Horseshoe Canyon, Royal Tyrrell Museum of Palaeontology, Drumheller, Alberta
4. Fossils of the Burgess Shale, Yoho National Park, Field, British Columbia
5. Precambrian Stromatolite Localities, Waterton-Glacier International Peace Park, Waterton Park, Alberta and West Glacier, Montana
6. Ginkgo Petrified Forest State Park, Vantage, Washington
7. John Day Fossil Beds National Monument, John Day, Oregon
8. Fossil Lake, Christmas Valley, Oregon
9. Hagerman Fossil Beds National Monument, Hagerman, Idaho
10. Sand Point Fossil Area, Hammett, Idaho
11. Malm Gulch Fossil Wood Area, Challis, Idaho
12. Dinosaur Nest Sites of the Willow Creek Anticline, Choteau, Montana
13. Petrified Forests, Yellowstone National Park, Mammoth Hot Springs Junction, Wyoming
14. Minnetonka Cave, Caribou National Forest, Saint Charles, Idaho
15. Fossil Butte National Monument, Kemmerer, Wyoming
16. Petrified Forest, Theodore Roosevelt National Park, Medora, North Dakota
17. Badlands National Park, Interior, South Dakota
18. The Mammoth Site of Hot Springs, South Dakota, Inc., Hot Springs, South Dakota
19. Agate Fossil Beds National Monument, Harrison, Nebraska
20. Kremmling Cretaceous Ammonite Locality, Kremmling, Colorado
21. Dinosaur Ridge, Morrison, Colorado
22. Florissant Fossil Beds National Monument, Florissant, Colorado
23. Garden Park Fossil Area, Cañon City, Colorado
24. Indian Springs Trace Fossil Site, Cañon City, Colorado
25. Clayton Lake State Park, Clayton, New Mexico
26. Blackwater Draw, Portales, New Mexico
27. Lubbock Lake Landmark, Lubbock, Texas
28. Dinosaur Valley State Park, Glen Rose, Texas
29. Big Bend National Park, Marathon, Texas
30. Capitan Reef, Guadalupe Mountains National Park, Pine Springs, Texas
31. Ghost Ranch *Coelophysis* Quarry, Ghost Ranch Conference Center, Abiquiu, New Mexico
32. Petrified Forest National Park, Holbrook, Arizona
33. Grand Canyon National Park, Grand Canyon Village, Arizona
34. Escalante Petrified Forest State Park, Escalante, Utah
35. Moab Dinosaur Tracks, Moab, Utah
36. Mill Canyon Dinosaur Trail, Moab, Utah
37. Dinosaur Hill/Riggs Hill Trails, Grand Junction, Colorado
38. Rabbit Valley Trail Through Time, Mesa County, Colorado
39. Cleveland-Lloyd Dinosaur Quarry, Price, Utah
40. Dinosaur National Monument, Jensen, Utah
41. Berlin-Ichthyosaur State Park, Austin, Nevada
42. Stewart Valley Paleontological Area, Hawthorne, Nevada
43. Hancock Park, Los Angeles, California
44. Ralph B. Clark Regional Park, Buena Park, California
45. Anza-Borrego Desert State Park, Borrego Springs, California
*Ashfall Fossil Beds State Historical Park, Royal, Nebraska

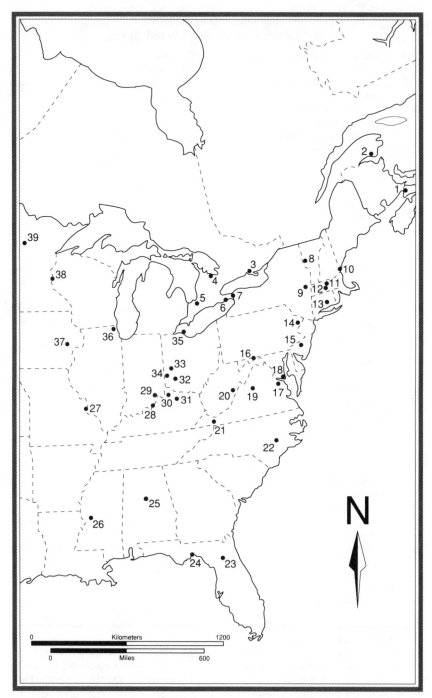

Figure 16. Map of eastern Canada and United States showing the location of interpreted fossil localities documented in J. N. McDonald, 1992, *Old Bones and Serpent Stones—Volume 1: Eastern Sites*. The sites are identified on the facing page.

Interpreted Fossil Localities
in Eastern Canada and United States

1. Joggins Cliffs, Joggins, Nova Scotia
2. Parc de Miguasha, Nouvelle, Quebec
3. Presqu'ile Provincial Park, Brighton, Ontario
4. Craigleith Provincial Park, Collingwood, Ontario
5. Rock Glen Conservation Area, Arkona, Ontario
6. Rock Point Provincial Park, Dunnville, Ontario
7. Niagara Gorge, Ontario and New York
8. Button Bay State Park, Vergennes, Vermont
9. John Boyd Thacher State Park, Voorheesville, New York
10. Odiorne Point State Park, Rye, New Hampshire
11. Barton Cove Natural Area, Gill, Massachusetts
12. Dinosaur Footprints Reservation, Holyoke, Massachusetts
13. Dinosaur State Park, Rocky Hill, Connecticut
14. Delaware Water Gap National Recreation Area, Pennsylvania and New Jersey
15. Pennypacker Park Dinosaur Monument, Cherry Hill and Haddonfield, New Jersey
16. Sideling Hill Road Cut and Visitor Center, Hancock, Maryland
17. Calvert Cliffs, Calvert County, Maryland
18. Westmoreland State Park, Montross, Virginia
19. Shenandoah National Park, Luray, Virginia
20. Cranberry Glades Botanical Area, Mill Point, West Virginia
21. Saltville Valley, Saltville, Virginia
22. Cliffs of the Neuse State Park, Seven Springs, North Carolina
23. Devils Millhopper State Geological Site, Gainesville, Florida
24. Edward Ball Wakulla Springs State Park, Wakulla Springs, Florida
25. Red Mountain Expressway Cut, Birmingham, Alabama
26. Mississippi Petrified Forest, Flora, Mississippi
27. Mastodon State Park, Imperial, Missouri
28. Falls of the Ohio Wildlife Conservation Area, Clarksville, Indiana and Louisville, Kentucky
29. Clifty Falls State Park, Madison, Indiana
30. Big Bone Lick State Park, Union, Kentucky
31. Blue Licks Battlefield State Park, Mount Olivet, Kentucky
32. Caesar Creek Dam Spillway, Waynesville, Ohio
33. Hueston Woods State Park, College Corner, Ohio
34. Aullwood Geology Trail, Dayton, Ohio
35. Glacial Grooves State Memorial, Kelleys Island, Ohio
36. Volo Bog State Natural Area, Ingleside, Illinois
37. Merrill A. Stainbrook Geological Preserve, North Liberty, Iowa
38. Interstate State Park, Taylors Falls, Minnesota and Saint Croix Falls, Wisconsin
39. Itasca State Park, Lake Itasca, Minnesota

many significant events in the evolution of life. For those readers interested in an evolutionary perspective, a list of the sites in order according to their ages, oldest to youngest, is provided (Table 1). There are some notable gaps in the fossil record as it is represented by public sites in western North America; some of the gaps in the western record are represented by public sites in eastern North America and are documented in *Old Bones and Serpent Stones—Volume 1: Eastern Sites* (Figure 16, Table 2).

Table 1. Fossil sites in western Canada and United States that are interpreted and open to public visitation. The sites are listed according to their geological age, from oldest to youngest, to facilitate reading from a paleontological perspective. Grand Canyon National Park is listed first because the concept of geological time is best illustrated by the rocks exposed there.

Site Number	Site Name	Geological Age	Page
33	Grand Canyon National Park	Precambrian to Cenozoic, with gap	217
5	Precambrian Stromatolite Localities	Middle Proterozoic	101
4	Fossils of the Burgess Shale	Middle Cambrian	94
24	Indian Springs Trace Fossil Site	Middle Ordovician	182
1	Stonewall Quarry Park	Late Ordovician	79
14	Minnetonka Cave	Early Carboniferous	137
30	Capitan Reef	Permian	203
31	Ghost Ranch *Coelophysis* Quarry	Late Triassic	209
32	Petrified Forest National Park	Late Triassic	213
41	Berlin-Ichthyosaur State Park	Late Triassic	248
35	Moab Dinosaur Tracks	Early Jurassic	226
23	Garden Park Fossil Area	Late Jurassic	178
39	Cleveland-Lloyd Dinosaur Quarry	Late Jurassic	238
40	Dinosaur National Monument	Late Jurassic	243
34	Escalante Petrified Forest State Park	Late Jurassic	223
36	Mill Canyon Dinosaur Trail	Late Jurassic	229
37	Dinosaur Hill/Riggs Hill Trails	Late Jurassic	232
38	Rabbit Valley Trail Through Time	Late Jurassic	235
21	Dinosaur Ridge	Early Cretaceous	169
25	Clayton Lake State Park	Early Cretaceous	186
28	Dinosaur Valley State Park	Early Cretaceous	195
2	Dinosaur Provincial Park	Late Cretaceous	85
3	Dinosaur Trail/Horseshoe Canyon	Late Cretaceous	90
12	Dinosaur Nest Sites of the Willow Creek Anticline	Late Cretaceous	129
20	Kremmling Cretaceous Ammonite Locality	Late Cretaceous	164
16	Petrified Forest, Theodore Roosevelt National Park	Paleocene	146
7	John Day Fossil Beds National Monument	Eocene to Miocene	110
11	Malm Gulch Fossil Wood Area	Eocene	126
13	Petrified Forests, Yellowstone National Park	Eocene	133
15	Fossil Butte National Monument	Eocene	141
29	Big Bend National Park	Eocene	199
17	Badlands National Park	Oligocene	151
22	Florissant Fossil Beds National Monument	Oligocene	174
6	Ginkgo Petrified Forest State Park	Miocene	105

19	Agate Fossil Beds National Monument	Miocene	160
42	Stewart Valley Paleontological Area	Miocene	252
*	Ashfall Fossil Beds State Historical Park	Miocene	265
9	Hagerman Fossil Beds National Monument	Pliocene	118
10	Sand Point Fossil Area	Pliocene	123
45	Anza-Borrego Desert State Park	Pliocene to Pleistocene	262
44	Ralph B. Clark Regional Park	Pleistocene	259
43	Hancock Park	Pleistocene	255
8	Fossil Lake	Pleistocene	114
18	The Mammoth Site of Hot Springs, South Dakota, Inc.	Late Pleistocene	155
26	Blackwater Draw	Late Pleistocene	189
27	Lubbock Lake Landmark	Late Pleistocene	192

Table 2. Fossil sites in eastern Canada and United States that are interpreted and open to public visitation. The sites are listed according to their geological age, from oldest to youngest, to facilitate reading from a paleontological perspective. The sites are described in Jerry N. McDonald, 1992, *Old Bones and Serpent Stones—Volume 1: Eastern Sites.*

Site Number	Site Name	Geological Age
19	Shenandoah National Park	Cambrian
25	Red Mountain Expressway Cut	Cambrian to Carboniferous
38	Interstate State Park	Cambrian; Late Quaternary
3	Presqu'ile Provincial Park	Ordovician
4	Craigleith Provincial Park	Ordovician
32	Caesar Creek Dam Spillway	Ordovician
34	Aullwood Geology Trail	Ordovician
7	Niagara Gorge	Ordovician to Silurian
29	Clifty Falls State Park	Ordovician to Silurian
9	John Boyd Thacher State Park	Ordovician to Devonian
8	Button Bay State Park	Ordovician; Late Quaternary
33	Hueston Woods State Park	Ordovician; Late Quaternary
14	Delaware Water Gap National Recreation Area	Silurian to Devonian
2	Parc de Miguasha	Devonian

(Table 2, continued)

5	Rock Glen Conservation Area	Devonian
6	Rock Point Provincial Park	Devonian
28	Falls of the Ohio Wildlife Conversation Area	Devonian
35	Glacial Grooves State Memorial	Devonian
37	Merrill A. Stainbrook Geological Preserve	Devonian
1	Joggins Cliffs	Carboniferous
16	Sideling Hill Road Cut and Visitor Center	Carboniferous
11	Barton Cove Natural Area	Triassic
12	Dinosaur Footprints Reservation	Triassic
13	Dinosaur State Park	Jurassic
15	Pennypacker Park Dinosaur Monument	Cretaceous
22	Cliffs of the Neuse State Park	Cretaceous; Eocene; Quaternary
23	Devils Millhopper State Geological Site	Eocene to Miocene; Quaternary
26	Mississippi Petrified Forest	Oligocene
17	Calvert Cliffs	Miocene
18	Westmoreland State Park	Miocene to Pliocene
10	Odiorne Point State Park	Late Quaternary
20	Cranberry Glades Botanical Area	Late Quaternary
21	Saltville Valley	Late Quaternary
24	Edward Ball Wakulla Springs State Park	Late Quaternary
27	Mastodon State Park	Late Quaternary
30	Big Bone Lick State Park	Late Quaternary
31	Blue Licks Battlefield State Park	Late Quaternary
36	Volo Bog State Natural Area	Late Quaternary
39	Itasca State Park	Late Quaternary

Note: The Quaternary Period spans the past 1.6 million years and includes the Pleistocene and Holocene epochs.

1. Stonewall Quarry Park

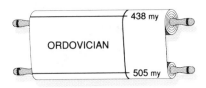

ORDOVICIAN

438 my

505 my

Stonewall, Manitoba

Limited outcroppings of Late Ordovician carbonate rocks, both limestones and dolomites, in central Manitoba attest to the presence, 450 million years ago, of a vast epicontinental sea flooding the interior of proto-North America. At that time, the continent straddled the equator, which extended diagonally across what is now northern North America (Figure 17); and the waters of the epicontinental sea were warm and tropical, the conditions ideal for the prolific growth of many marine organisms.

The interior of North America during the Paleozoic exhibited moderate topographic relief as a series of basins and arches developed in the Precambrian basement rock, structural depressions and highlands that played a significant role in controlling the distribution of organisms living in the epicontinental sea. The Williston Basin, centered in southern Saskatchewan and the Dakotas, was one of the largest depressions. On its margins, reef-like mounds of tabulate corals, stromatoporoids, and bryozoans grew; beyond the basin, the waters of the epicontinental sea were often shallow enough that the sea floor was constantly within the zone of wave agitation, conditions conducive to animal growth but not ideal for the preservation of fossils. As a result, the shells of most of the animals were broken up and eventually pulverized into lime muds by the relatively high energy of the water. Reef-like mounds were also scattered on the sea floor, and among them, there was shelter for a variety of animals. In such areas, horn corals, brachiopods, nautiloids, and pieces of crinoids were preserved intact. From time to time, however, the waters of the epicontinental sea became shallow, the reef-like mounds restricted water circulation, and the water became saline; the number of organisms that could survive such conditions decreased dramatically, and evaporite minerals such as gypsum and halite were precipitated.

The rocks exposed at Stonewall Quarry Park (Stonewall and Stony Mountain formations) are a small portion of an extensive sheet of buried carbonates that extend in a broad swath from Nevada and Arizona in

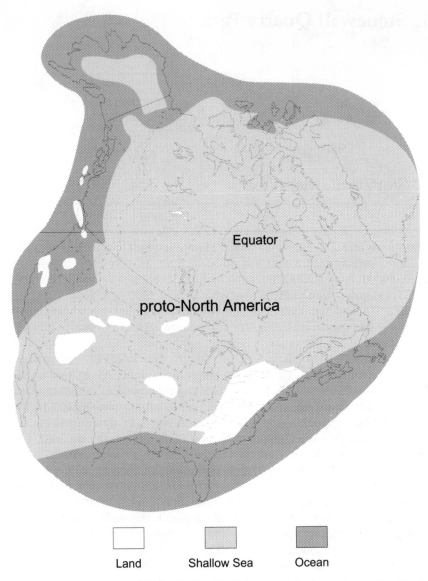

Land   Shallow Sea   Ocean

Figure 17. Paleogeography of North America in Late Ordovician time. Proto-North America was almost entirely flooded by shallow epicontinental seas.

the southwest northeastward into the interlake area of south-central Manitoba (Figure 18). The greatest lateral extent of the limestones occurs within the Williston Basin. Similar carbonate deposits are known farther to the northeast (in Hudson Bay and on Baffin Island) and to the north (in Yukon and the Canadian Arctic Islands). It is reasonable to infer that the sheets of limestone and dolomite existing today were probably once

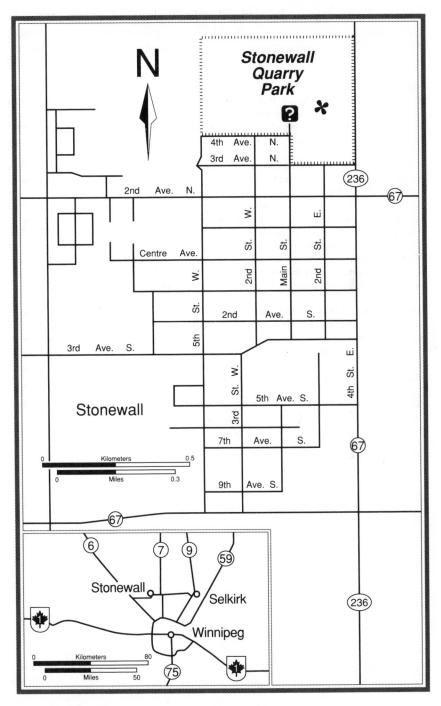

Figure 20. Location of Stonewall Quarry Park, Stonewall, Manitoba.

industrial lime. Stonewall Quarry Park highlights the latter use; the park itself, situated on the site of an old quarry, preserves many historical structures. The geological exposures are limited to old, open-pit quarry workings.

DIRECTIONS: Stonewall Quarry Park is located within the town of Stonewall in the inter-lake area of south-central Manitoba north of Winnipeg. Access to the park is from Provincial routes 7 and 67. From the intersection of the two highways follow Provincial Route 67 west for 5.8 kilometers (3.4 miles) to Main Street; proceed north on Main Street 0.3 kilometer (0.2 mile) to the park entrance and parking area (Figure 20).

PUBLIC USE: Season and hours: Stonewall Quarry Park, owned and operated by the town of Stonewall, is open year round: daily in summer from 9:00 A.M. to 9:00 P.M.; Tuesday to Sunday in winter from 9:00 A.M. to 5:00 P.M. **Fees:** $3.00/person daily except $2.00/person for students and individuals over 65 years of age, children 5 years and younger admitted free. Reduced rates apply in winter. **Food service:** Restaurants and stores are available in Stonewall, and a food concession is operated within the park. **Recreational activities:** The range of recreational activities includes swimming, picnicking, baseball, and camping (for which a separate fee is charged, but it includes the $3.00/person admission fee). There is also a playground area. **Handicapped facilities:** The Visitor Reception Centre is accessible by wheelchair. **Restrictions:** Collecting of fossils is prohibited.

EDUCATIONAL FACILITIES: Visitor Center: A Visitor Reception Centre features geologi-cal and paleontological, as well as historical displays. A short slide presentation documents the history of the area, and fossils are available for examination. **Visitor Center hours:** Same as park hours of operation. **Fees:** None. **Bookstore:** A small bookstore and gift shop stocks articles of local interest. **Trails:** A short, easy-to-walk, self-guiding trail, which highlights both fossil and historical aspects of the park, takes visitors around the old quarry.

FOR ADDITIONAL INFORMATION: Contact: The Manager, Stonewall Quarry Park, P.O. Box 250, Stonewall, Manitoba R0C 2Z0, (204) 467–5354.

2. Dinosaur Provincial Park

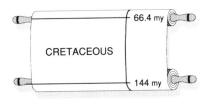

CRETACEOUS

66.4 my

144 my

Brooks, Alberta

In Late Cretaceous time, the western interior of North America was flooded from the Gulf of Mexico to the Beaufort Sea by the Western Interior Seaway, an epicontinental sea which fluctuated in size as sea levels world-wide rose and fell. Some 75 million years ago, central and eastern Alberta was part of a broad coastal plain that had developed on the western margin of the Western Interior Seaway (Figure 21). The Cordilleran Mountains to the west were young and active, producing a constant supply of sediment to the rivers that drained the mountains and flowed into the epicontinental seaway. The climate in the Late Cretaceous on the margin of the Western Interior Seaway was subtropical and humid. The landscape on the coastal plain was flat and featureless, a broad lowland characterized by sluggish meandering rivers, wide and well-vegetated floodplains, marshes and oxbow lakes, and estuaries near the shore of the epicontinental sea. Many volcanoes were active during that time in the mountains to the west, and the ashfalls they produced greatly enriched the heavy clay soils of the coastal plain. The result was a lush and diverse flora, which supported a large and complex association of animals. Dinosaurs, the prominent animals, are represented by 36 species; but mammals, too, were a significant component of the terrestrial fauna, and 20 species are known. Large herds of dinosaurs lived on the coastal plain or migrated through, feeding on the lush vegetation of the lowland swamps and river valleys. The most common among them were the duckbilled dinosaurs (hadrosaurs), such as *Corythosaurus* and *Lambeosaurus*, and rhinoceros-like horned dinosaurs, such as *Centrosaurus* and *Styracosaurus*. The most common large carnivore was the tyranosaurid, *Albertosaurus*. Pachycephalosaurs were the smaller herbivorous dinosaurs, recognized by their characteristic but peculiar cranial domes. *Troodon* was a small, sophisticated carnivore and scavenger with large brain and huge eyes. Pterosaurs were visible in the air; lizards, crocodiles, and turtles lived in the swamps and lakes of the coastal plain.

The dinosaur bones were preserved in ways that are almost as varied as the dinosaurs themselves. Isolated articulated skeletons represent ani-

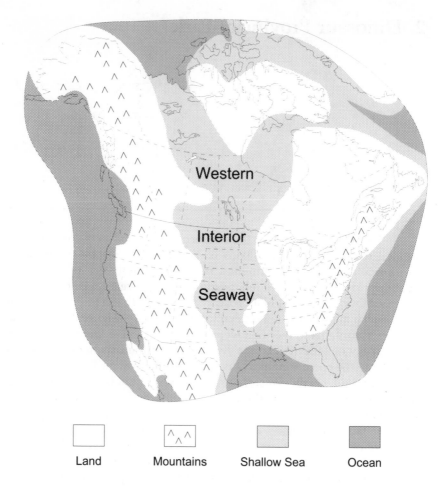

Land	Mountains	Shallow Sea	Ocean

Figure 21. The Western Interior Seaway during Late Cretaceous time.

mal carcasses that fell in the river, floated downstream, and eventually became lodged, perhaps attacked by scavengers, and buried. Sometimes complete, but frequently not, these fossils are unusual finds, but they are not rare. Much more common are isolated bones, some the disarticulated remains of animals that died at a distance away from the river's edge, the carcasses attacked by both scavengers and the elements, and ultimately scattered on the ground and buried.

Bonebeds, mass accumulations of disarticulated bones, are common in Dinosaur Provincial Park. The best known among them are the *Centrosaurus* Bonebed and the *Styracosaurus* Bonebed. Such mass accumulations typically represent a mass mortality either of a herd of animals (a single species) or a community of animals (several species) and can be accounted for in several ways by analogy with modern events. Perhaps drought forced animals to congregate at a persistent watering hole; when the water disappeared, animals died and the carcasses accumulated. Perhaps a flash flood trapped a herd of animals and swept them downstream. In either case, carcasses were collected by the moving water and carried downstream *en masse*. If the water levels were high, the river deposited the carcasses high on point bars and stranded them there, to be attacked by scavengers as flood waters receded. Later, as the point bar was undercut by the flow of water, the bones fell into the stream where they were buried and preserved.

Sites with unusually high concentrations of bones and teeth of small animals are also abundant in Dinosaur Provincial Park. Not as visually impressive as bonebeds, they are nonetheless important because they reveal the small animal component of the ecosystem, a component that at Dinosaur Provincial Park is largely made up of mammals. Thus a more complete view of life in the past is possible.

The sediments of the Judith River Formation, deposited during Late Cretaceous time in what is now central Alberta, are widely exposed in the Red Deer River Badlands of Alberta. The badlands topography exposes the sediments horizontally for many kilometers and vertically to a thickness of 90 meters (300 feet). The sediments are variegated gray, buff, and pink silts and clays with channel sandstones, shales, and ironstones. They are horizontally bedded and laterally continuous, overlain by the swamp sediments and coal seams that mark the transition to the Bearpaw Shale, a marine deposit that records the last expansion of the Western Interior Seaway.

Dinosaur Provincial Park protects the accumulation of dinosaurs in the Judith River Formation, one of the most extensive known anywhere in the world. The park itself is divided into two areas: a public use area to which there is unrestricted access and a Natural Preserve which is closed to the public except by guided tour. The Natural Preserve was set aside to protect the fossils and to facilitate research. Although dinosaur bone fragments are not as common in the heavily used areas of the park as in the Natural Preserve, fossilized material can still be seen. Meanwhile, erosion continues to expose new fossils, and hikers may make a fortuitous discovery. The unique attributes of the park were acknowledged when the park was declared a UNESCO World Heritage Site.

DIRECTIONS: Dinosaur Provincial Park is located in a sparsely populated portion of Alberta's ranching country. Most direct access is from Brooks on the TransCanada Highway. The park is approximately 45 kilometers (27 miles) northwest of Brooks. The route to the park, via secondary paved roads, is posted. From its junction with the TransCanada Highway, follow Provincial Route 873 north for 10.3 kilometers (6.4 miles), then proceed east along Provincial Route 544 for 16 kilometers (10 miles). The town of Patricia is located just north of the intersection. Follow the paved road north past Patricia 3.2 kilometers (2 miles) to a T-junction, then east as the road winds eastward and northward 13.3 kilometers (8.2 miles) to the park entrance. The Royal Tyrrell Museum of Palaeontology Field Station is 1.2 kilometers (0.7 mile) beyond the entrance (Figure 22).

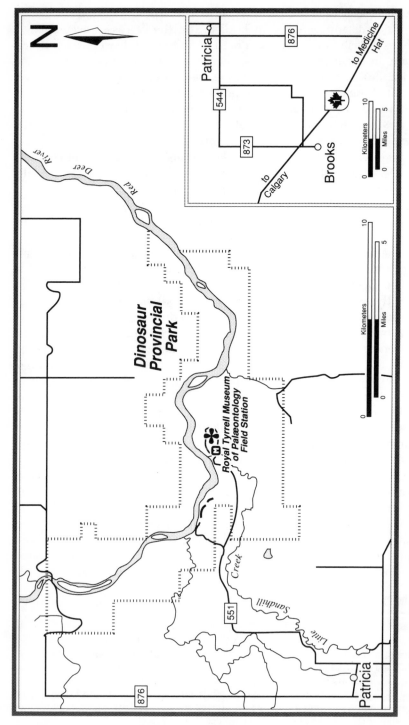

Figure 22. Location of Dinosaur Provincial Park, Alberta.

An alternative route may appeal to some travellers: from the TransCanada Highway at Suffield, follow Provincial Route 884 north past Canadian Forces Base Suffield to the junction with Provincial Route 544, then east to Patricia to join the main access route. The distance is slightly shorter and the roads less heavily travelled. Approaching the park from other directions is not recommended because the roads are secondary and unpaved, and the conditions are variable. For some travellers, however, the saving in distance may be worth the risks. Those travellers should obtain a good, large-scale map of grid roads in Alberta.

PUBLIC USE: Season and hours: Dinosaur Provincial Park is open to the public year round, but outdoor activities may be restricted by winter conditions. **Fees:** None. **Food service:** Food services are limited. A small store is operated within the park during summer months, and several small variety stores are present along the route to the park (hours of operation variable). **Recreational activities:** A variety of recreational activities is available including camping (for which a fee is charged), hiking, canoeing, and fishing. Wildlife is abundant within the park. **Handicapped facilities:** The Royal Tyrrell Museum of Palaeontology Field Station and one of the outdoor dinosaur exhibits along the loop road through the park are accessible by wheelchair. **Restrictions:** Collecting of fossils is prohibited.

EDUCATIONAL FACILITIES: Visitor Center: The Royal Tyrrell Museum of Palaeontology Field Station operates as a visitor center and a field research station. It offers a field laboratory setting within which visitors can see scientists at work and view a fossil preparation laboratory. It includes displays, videos, computer simulations, and a theater. The interpretation that it offers concentrates on the results of research conducted within the park. **Visitor Center hours:** The Field Station is open year round: daily in summer, May (Victoria Day weekend) to October (Thanksgiving Day), from 9:00 A.M. to 9:00 P.M.; Wednesday to Sunday in winter from 10:00 A.M. to 5:00 P.M. (or by appointment); except closed on Christmas Day (December 25). **Fees:** None. **Trails:** A 3.5 kilometer (2.2 mile) loop road allows visitors to drive through the public use area of the park. Along that route are two interpreted nature trails; of geological interest is the Badlands Interpretive Trail, self-guiding, short (1.5 kilometers or 1 mile) and easy to walk. There are two sheltered dinosaur displays along the eastern portion of the loop road which can serve as starting points for other, more strenuous, but not interpreted, trails into the badlands. **Special group activities:** Special tours can be arranged for school groups and commercial tour groups. **Staff programs:** Interpretation programs include guided hikes, narrated bus tours to the Natural Preserve, visits to a bonebed excavation, and evening amphitheater programs on weekends. **Note:** A volunteer program operated from the Royal Tyrrell Museum of Palaeontology in Drumheller allows non-paleontologists to spend time in the field digging for dinosaurs.

FOR ADDITIONAL INFORMATION: Contact: Chief Park Ranger, Dinosaur Provincial Park, Box 60, Patricia, Alberta T0J 2K0, (403) 378–4587. **Read:** (1) Danis, Jane. 1988. Bibliography of vertebrate palaeontology in Dinosaur Provincial Park. Alberta: Studies in the Arts and Sciences, volume 1, number 1, 225–234 pp. (2) Koster, Emlyn H. 1987. Vertebrate taphonomy applied to the analysis of ancient fluvial systems. Pp. 159–168 in Ethridge, Frank G., Romeo M. Flores, and Michael D. Harvey (editors). Recent Developments in Fluvial Sedimentology. Society of Economic Paleontologists and Mineralogists, Special Publication Number 39. (3) Gross, Renie. 1985. Dinosaur Country: Unearthing the Badlands' Prehistoric Past. Saskatoon, Saskatchewan: Western Producer Prairie Books.

3. Dinosaur Trail/Horseshoe Canyon
Royal Tyrrell Museum of Palaeontology

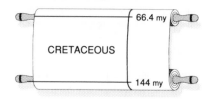

66.4 my

CRETACEOUS

144 my

Drumheller, Alberta

The first dinosaurs of scientific interest to be found in Canada were discovered by Joseph Burr Tyrrell. In 1884, Tyrrell and his survey crew from the Geological Survey of Canada were exploring the Red Deer River Badlands in the vicinity of what is now the city of Drumheller when they discovered their first dinosaur in sediments on the river bank. Today, the Royal Tyrrell Museum of Palaeontology commemorates both that discovery and the science of paleontology.

In the Late Cretaceous, dinosaurs lived in a lush, subtropical jungle on the western shores of the Western Interior Seaway. Only about 5 million years had passed since the dinosaurs of Dinosaur Provincial Park [Site 2] were preserved in sediments of the Judith River Formation. Herbivorous dinosaurs still browsed, 70 to 68 million years ago, among giant cypress trees and were stalked by carnivores, large and small; small, furry mammals scurried about underfoot. Some of the species were different, but little else had changed. Coals in the sediments attest to the presence of lush vegetation and swamps, bentonites (altered volcanic ash) to the continued activity of volcanoes to the west, and oyster beds to the abundance of water.

The Red Deer River Badlands at Drumheller are not as extensively developed and the Red Deer River not as deeply incised as at Dinosaur Provincial Park farther downstream. The sediments exposed at Drumheller, the Horseshoe Canyon Formation (or Edmonton Formation), stratigraphically overlie the sediments of the Judith River Formation and are, therefore, younger than the sediments at Dinosaur Provincial Park. The two formations are similar in appearance, however. Both consist of variegated gray, buff, and pink silts and clays, but abundant coal seams and indurated sandstone lenses present throughout the Horseshoe Canyon Formation indicate that the area was wetter and coal swamps were abundant. The coal seams are consistent with evidence of rising sea level in the Western Interior Seaway at that time, of the sea encroaching westward onto the land.

The Horseshoe Canyon Formation is overlain by yellow-orange clays that were deposited in a glacial lake as the last ice sheet melted away about 10,000 years ago. The unconformity, the gap in the geological record, represents more than 65 million years of earth history.

The Horseshoe Canyon Formation at Drumheller does not have the immense accumulation of dinosaur bones that the Judith River Formation has. Nonetheless, new fossils, usually occurring as isolated bones or as portions of skeletons, are continually being exposed by relentless erosion. Dinosaur bones may not always be visible, but evidence of past life can be seen everywhere. The coal seams record abundant plant life. Impressions of plants can be seen in the layers of sandstone, including those used as paving stone on the badlands trail at the museum. Even the concretions, the massive blocks of rusty-red rocks occurring in layers throughout the badlands, are evidence of past life. They mark ancient soils, the organic-rich sediments highlighted by the precipitation of iron and other minerals from the groundwater.

DIRECTIONS: The Royal Tyrrell Museum of Palaeontology, on the west side of Midland Provincial Park, is located in the badlands of the Red Deer River at Drumheller (Figure 23). Access is by way of North Dinosaur Trail (Provincial Route 838) 6 kilometers (3.5 miles) west of Provincial Route 9. Horsethief Canyon is 10 kilometers (6 miles) west of the museum on North Dinosaur Trail; Horseshoe Canyon, which provides ready access for exploring in the badlands, is located 19 kilometers (11 miles) west of Drumheller on Provincial Route 9.

PUBLIC USE: Season and hours: Midland Provincial Park, and Horsethief and Horseshoe canyons are open to the public year round. Access is limited by weather conditions. It is particularly important to note that the bentonite-rich clays of the badlands are extremely slippery and dangerous when wet. Under such conditions it may not be possible to climb slopes in the badlands or drive on undeveloped roads. **Fees:** None. **Food service:** The museum houses a cafeteria, and restaurants and stores are available in Drumheller. **Recreational activities:** Hiking and exploring in the badlands are popular activities within and beyond Midland Provincial Park. Self-guiding trails emphasize the history of coal mining in the region. Other recreational activities, such as camping, are available in the Drumheller area. **Handicapped facilities:** The Royal Tyrrell Museum of Palaeontology is accessible by wheelchair. **Restrictions:** Collecting of fossils is prohibited.

EDUCATIONAL FACILITIES: Museum: The Royal Tyrrell Museum of Palaeontology documents the history of life on earth and highlights the paleontology of Alberta, in particular the dinosaurs. By good fortune, much of geological time is represented by rocks in Alberta, and many are fossiliferous. Those fossils are the focus of exhibits at the museum. Mounted skeletons of many animals, especially local dinosaurs, are displayed. The museum also offers videos, computer puzzles, and a view of the fossil preparation laboratory. Of special interest is the palaeo-conservatory, which houses archaic plants whose ancient relatives shared the world with the dinosaurs. **Museum hours:** The Royal Tyrrell Museum of Palaeontology is open year round: daily in summer, May (Victoria Day weekend) to October (Thanksgiving Day), from 9:00 A.M. to 9:00 P.M.; Tuesday to Sunday in winter from 10:00 A.M. to 5:00 P.M. **Fees:** Adults $5.00, 7–17 years $2.00, under 7 years free; no fees charged on Tuesdays. **Bookstore:** A small bookstore and gift shop offers a variety of books and souvenirs, the emphasis on items that feature dinosaurs. **Trails:** A short, easy-to-walk, self-guiding trail leads into the badlands immediately adjacent to the museum. Fossils are not highlighted but the observant walker may spot a variety of fossils, especially of plants.

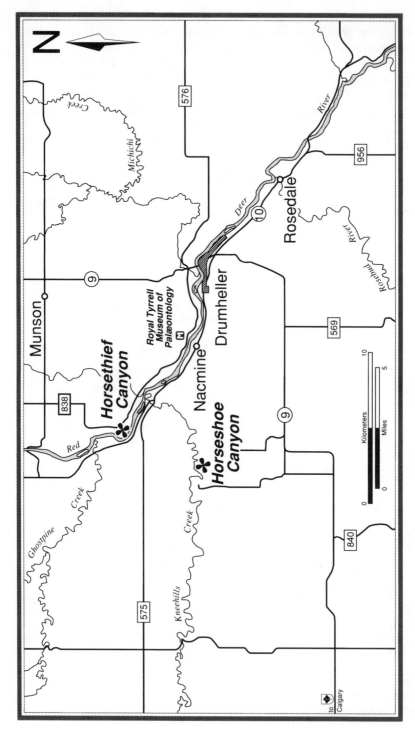

Figure 23. Location of Dinosaur Trail/Horseshoe Canyon, Royal Tyrrell Museum of Palaeontology, Alberta.

A trail from the parking area at Horsethief Canyon leads down the canyon wall to oyster beds; at Horseshoe Canyon, trails provide access for exploring the badlands. **Special group activities:** Tours for special interest groups, especially school groups, are available by prior arrangement. A variety of special events including travelling exhibits, public lectures, interpretive drama and games, audio-visual presentations, conferences, and contests are presented throughout the year.

FOR ADDITIONAL INFORMATION: Contact: Royal Tyrrell Museum of Palaeontology, Box 7500, Drumheller, Alberta T0J 0Y0, (403) 823–7707 (direct call from Calgary 294-1992). **Read:** (1) Foster, John, and Dick Harrison (editors). 1988. Tyrrell Museum of Palaeontology (a special commemorative issue featuring 20 separate articles). Alberta: Studies in the Arts and Sciences, volume 1, number 1. (2) Sternberg, Charles H. 1985. Hunting Dinosaurs in the Badlands of the Red Deer River, Alberta Canada. Third Edition, introduced by David A. E. Spalding. Edmonton, Alberta: NeWest Press.

4. Fossils of the Burgess Shale
Yoho National Park

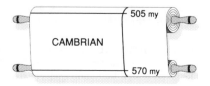

505 my

CAMBRIAN

570 my

Field, British Columbia

The fossils of the Burgess Shale are the most famous fossils in Canada. They were first discovered on Mt. Field in 1909 by Charles D. Walcott who recognized their unique aspect and collected them extensively. Walcott found spectacularly preserved soft-bodied animals, in unsuspected abundance and disparity. The significance of the fossils from the Burgess Shale is underscored by the fact that Yoho National Park and the fossil-bearing beds that it contains are designated a UNESCO World Heritage Site.

In modern environments, soft-bodied organisms are a significant component of life on earth, especially life in the oceans. Yet the fossil record preserves almost exclusively the history of animals with hard parts. Such bias makes it impossible to document the evolutionary history or the ecological and morphological diversity of soft-bodied animals. The fauna of the Burgess Shale, one of the vagaries of the fossil record, is one of the few that allows paleontologists to study the history of soft-bodied animals.

The main geological features of Yoho National Park that circumscribe the fossils of the Burgess Shale are exposed on the southeast face of Mt. Field. Three formations can be recognized: the lowest, the Cathedral Formation, is overlain by the Stephen Formation which includes the Burgess Shale, the sequence is capped by the Eldon Formation. The stratigraphic relationships reveal the unique sequence of events of Middle Cambrian time.

The Rocky Mountains of today did not exist 530 million years ago; instead what is now the continental divide marked the edge of a smaller continent, proto-North America. Ocean levels were rising, and the sea was flooding over onto land (Figure 24). Western North America lay in the tropics; the seas were warm, and ocean life was plentiful. On the edge of the continent, for a distance of hundreds of kilometers, grew calcareous algae, stromatolite-building cyanobacteria (so-called blue-green algae), and branching bryozoans. This was a zone of prolific organic growth, and eventually a great thickness of limestone accumu-

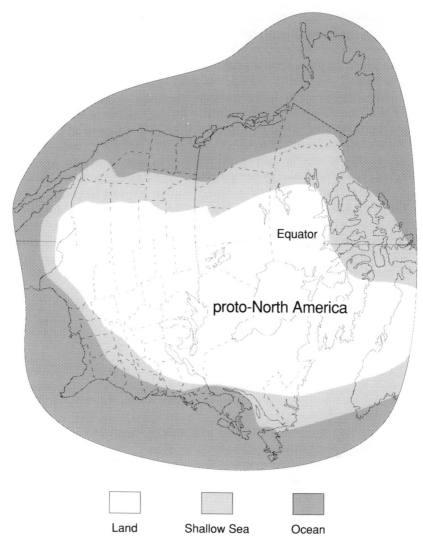

Land   Shallow Sea   Ocean

Figure 24. Paleogeography of proto-North America in Middle Cambrian time.

lated (Cathedral Formation). To the landward side, what is now toward the northeast, lime muds made mostly of finely pulverized fossil shells were deposited in the quiet, sheltered waters of the epicontinental sea. Oceanward, presently on the southwest, where the water was deep, shales and calcareous muds were deposited (Stephen Formation). After a time, the rise in sea level was sufficient that the entire area was flooded with deep water, and sediments of the Stephen Formation were blanketed over the Cathedral limestones. Eventually, the limestones of the Eldon Formation were deposited over the whole area. Today the rocks of the

Cathedral Formation, which are more resistant to erosion than the surrounding rocks, form a prominent escarpment.

The Burgess Shale occurs mid-way up in the thick sequence of sediments of the Stephen Formation along the edge of the escarpment, within 20 meters (65 feet) of the contact with the Cathedral Formation. It is a pod-like lens of sediment that was deposited when water-saturated muds and silts, accumulating on the gently sloping surface at the edge of the continent, became unstable and moved suddenly downslope as a mudflow, carrying organisms in suspension in the sediment. Sediments and fossils came to rest in deeper, anoxic waters where organic decay was limited. The spectacular preservation of soft-bodied organisms resulted.

Approximately 140 species of organisms, most of them animals, have been identified from the Burgess Shale (Figure 25). Most of the animals are arthropods, soft-bodied relatives of the much better known trilobites that flourished in the Cambrian oceans, ancient relatives of such animals as spiders and crabs familiar to us in the modern world. Worms, too, are well represented. Sponges, brachiopods, echinoderms, and mollusks are known. Significantly, there are also animals present in the Burgess Shale that are unlike any animals known today, animals whose graphic names betray their unknown affinities—*Anomalocaris, Hallucigenia, Wiwaxia*.

The fossils from the Burgess Shale highlight important characteristics of the beginning of modern life. For one, they show that modern animals were already well-established; all six modern body plans are known. Polychaete worms, for example, are represented by six families in the Burgess Shale; all families found as fossils survive to this day. Even chordates, the precursors to the vertebrates, are represented, *Pikaia* being identified as the first ancestral chordate.

The second characteristic was as surprising as it is revolutionary to preconceptions in paleontology. Among the Burgess animals are body plans that no longer exist: several flat, swimming animals such as *Odontogriphus* and *Amiskwia*; *Nectocaris*, part arthropod, part chordate, but mostly neither; a tiny wine-glass shaped animal named *Dinomischus*; bizarre *Hallucigenia*.

Equally revolutionary is the view of the ecology afforded by the Burgess fauna. It shows that complex community and ecological relationships were fully in place at the very earliest stages of the Cambrian Explosion. Of particular significance is that the first true predators that are known are found among the soft-bodied organisms of the Burgess Shale. Every niche that characterizes modern marine communities was occupied: sediment feeders lived on (*Marrella*, an arthropod) or within (*Ottoia*, a priapulid worm) the bottom sediments, free-swimming organisms (*Odontogriphus*) occupied the open waters, filter feeders (*Dinomischus*) were attached to objects on the sea floor, carnivores and scavengers were well represented, and true predators complete the picture of ecological

96

Figure 25. The fauna of the Burgess Shale (Smithsonian Institution Photograph No. SI–84–4711)

complexity (the largest, *Anomalocaris*, has no modern counterpart).

The lens of shale that yielded the soft-bodied organisms at Walcott's original quarry has been completely excavated, but other areas in the park where similar shale is exposed are also fossiliferous. Without the unique aspect of deposition that characterizes the Burgess Shale, however, most shales of the Stephen Formation contain only fossils of animals

97

with hard parts. One such outcrop, an old quarry in dark gray shales located on Mt. Stephen, is called the Trilobite Fossil Beds because the dominant fossils there are the trilobites. The preservation in most localities is such that *in situ* fossils are singularly not spectacular: the soft-bodied organisms occur as inky smears on black shale; trilobites, as black shapes in black shale. The fossils selected for display at indoor exhibits are much more impressive.

DIRECTIONS: Yoho National Park is located on the western slopes of the Continental Divide, the boundary between Yoho and Banff national parks. The park is bisected by the TransCanada Highway; the town of Field is located approximately in the center of the park (Figure 26).

PUBLIC USE: Season and hours: Yoho National Park is open to the public year round although most visitor services are available only in summer. Fossil sites in the park are closed to the public unless accompanied by uniformed park staff or licenced private guides. **Fees:** $4.00 daily, $9.00/4-day, or $25.00 annual Parks Canada Permit. **Food service:** Restaurants and stores are available in Field, and there is a convenience store on Yoho Valley Road (Takakkaw Falls Road) across from the entrance to Kicking Horse Campground. **Recreational activities:** A full range of park activities and facilities is available including camping (for which an additional fee is charged), hiking trails, Hostel, backpacking (a free permit is required for back-country camping), fishing (licence required), mountain biking (in designated areas only), and winter activities such as cross-country skiing and ski touring, ice-climbing, and winter camping. Private lodges, commercial services (including rental of equipment), and recreation concessions (such as horseback riding) are available, some remaining open in winter. **Restrictions:** Collecting of fossils is prohibited.

EDUCATIONAL FACILITIES: Visitor Center: The park Information Centre in Field is the main point of contact between visitors and park staff, functioning primarily as a source of information but also housing an exhibit of fossils from the Burgess Shale. Additional interpretive information on the geological history of the Rocky Mountains and on the fossils of the Burgess Shale is presented at the Lake Louise Visitor Reception Centre in adjacent Banff National Park. **Visitor Center hours:** The Information Centre is open year round: daily; Monday to Thursday from 8:00 A.M. to 4:00 P.M., Friday to Sunday from 8:00 A.M. to 5:30 P.M.; with extended hours mid-June to Labor Day to 9:00 P.M., and Labor Day to October 1 to 6:00 P.M. **Fees:** None. **Bookstore:** The sales outlet operated by the Friends of Yoho offers items such as books, maps, and calendars, and a limited number of publications on the fossils from the Burgess Shale. **Interpretive sign:** Interpretive panels depicting the Burgess Shale, its fauna, and the history of the site are exhibited at the Kicking Horse Overflow Campground located 1 kilometer (0.6 mile) from the TransCanada Highway on Yoho Valley Road. **Staff programs:** Interpretive programs, campfire and theater presentations, and guided hikes are available in the summer (schedule of events available at information centers). Public access to fossil sites is available to visitors only by joining park interpreters on regularly scheduled guided hikes during July and August to the Trilobite Fossil Beds on Mt. Stephen (on Monday, Tuesday, Thursday, and Friday) and to the Walcott Quarry on Mt. Field (on Saturday). Participants must register in advance. Some reservations (for up to 5 people) are accepted by the Information Centre (phone (604) 343–6324, extension 10 or 24). Groups of 6–15 people can arrange for special presentations; during July and August, however, such requests are referred to licenced interpretive guides who will charge a fee for their services. Note that all hikes are limited to 15 people. The hikes to the Trilobite Fossil Beds and to the Walcott Quarry follow strenuous, short, steep trails that lead part way up Mt. Stephen and Mt. Field. At the quarries, on the scree slope, fossils can be seen on the broken slabs of split shales.

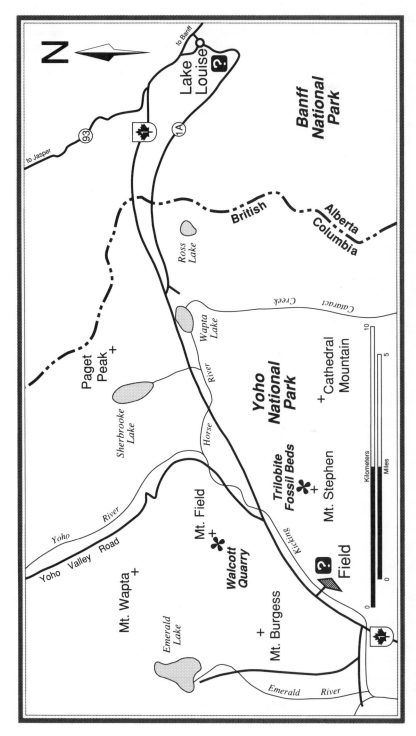

Figure 26. Location of Field in Yoho National Park, British Columbia, for access to the fossils of the Burgess Shale.

FOR ADDITIONAL INFORMATION: Contact: Superintendent, Yoho National Park, P.O. Box 99, Field, British Columbia, V0A 1G0, (604) 343–6324. **Read:** (1) Conway Morris, Simon (editor). 1982. Atlas of the Burgess Shale. London, England: Palaeontological Association. (2) Conway Morris, Simon, and Harry B. Whittington. 1985. Fossils of the Burgess Shale: A National Treasure in Yoho National Park, British Columbia. Geological Survey of Canada, Miscellaneous Report 43. (3) Gould, Stephen Jay. 1989. Wonderful Life: The Burgess Shale and the Nature of History. New York, New York; London, England: W. W. Norton and Company. (4) Ludvigsen, Rolf. 1989. The Burgess Shale: not in the shadow of the Cathedral Escarpment. Geoscience Canada, volume 16, number 2, pp. 51–59. (5) Whittington, Harry B. 1985. The Burgess Shale. New Haven, Connecticut: Yale University Press.

5. Precambrian Stromatolite Localities
Waterton-Glacier International Peace Park

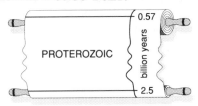

Waterton Park, Alberta
West Glacier, Montana

Stromatolites—large, three-dimensional structures in carbonate rocks; dome, mound, or pillar shapes with internal layering. They are the evidence that life on earth was widespread very early in earth history. Stromatolites, the oldest known are from rocks in Australia that are approximately 3.5 billion years old, became abundant in rocks about 2.5 billion years old (early Proterozoic). They remained the primary form of life on earth for almost 2.0 billion years throughout Proterozoic time, producing extensive reef-like limestone deposits, only becoming rare about 500 million years ago in Ordovician time.

Stromatolites are the products of so-called simple organisms, bacteria and cyanobacteria, living together in complex reef-like communities. These communities thrived in the shallow warm seas on the margins of nascent continents, the microbes growing upward toward the sunlight and manufacturing their food through photosynthesis. In the process, by extracting carbon dioxide from the atmosphere and releasing free oxygen into it, they altered the composition of the atmosphere. The carbon dioxide was stored in the form of calcium carbonate precipitated by the organisms on the sea floor. Thus bacteria and cyanobacteria inadvertently built massive carbonate mounds and relentlessly altered the world in which they lived: these simple organisms determined that the subsequent life on earth would be aerobic.

The rocks exposed in Waterton-Glacier International Peace Park were deposited in a subsiding ocean basin that flanked the proto-North American continent; not for another 500 or 600 million years would mountains begin to form here and the modern continent take shape. Lime muds and debris from the continent accumulated slowly; as the ocean basin sank, sediments accumulated to a total thickness of some 5500 meters (18,000 feet). They have since been lithified and metamorphosed, altered by high temperature and intense pressure; they have been folded, faulted, and uplifted to form mountains. Today those ancient sediments are known as the Belt-Purcell Supergroup. Dramatic evidence of mountain-building events is recorded by a feature called the Lewis

Thrust Fault, visible in sharp relief along the eastern boundary of the park. Here it is clear that the mountains of Waterton-Glacier International Peace Park (rocks that are about one billion years old) rest on top of rocks that are less than 100 million years old. Earth movements uplifted a giant slab of ancient rock and pushed it overtop of younger rock. As a result, ancient stromatolites that were once deeply buried have been exposed to view.

Stromatolite-bearing rocks, equivalent to those of the Belt-Purcell Supergroup, are exposed at the base of the Grand Canyon [Site 33]. There they are called the Grand Canyon Supergroup. Both sets of rocks represent laterally extensive deposition on the margin of the ancestral continent, but the post-depositional history of each is distinct.

Stromatolites that grew one billion years ago can be seen today in Waterton-Glacier International Peace Park and beyond in widely distributed carbonate rocks of the Belt-Purcell Supergroup. The fossils are most abundant and readily accessible in the Siyeh (in Canada) or Helena Formation (in United States), but they also occur in other rocks: in the Altyn Formation which is older and occurs only in the eastern-most portion of the park, and in the Snowslip Formation which overlies the Helena Formation.

DIRECTIONS: Waterton-Glacier International Peace Park, comprising Waterton Lakes National Park in Alberta and Glacier National Park in Montana, straddles the Canada-United States border and the continental divide in the United States. There are a number of access routes to the park but the primary ones are Alberta Provincial routes 5, 6, and 3, and United States routes 2 and 89 (Figure 27). Travel between the parks via Chief Mountain International Highway (State Route 17, Provincial Route 6) is possible only in the summer (customs offices open daily mid-May, exact date varies from year to year, to May 31 from 9:00 A.M. to 6:00 P.M.; June 1 to September 14 from 7:00 A.M. to 10:00 P.M.). Access to fossil exposures is by car along Going-to-the-Sun Road (Glacier National Park), by hiking from Red Rock Canyon to Goat Lake (Waterton Lakes National Park), or by canoe from Cameron Lake (Waterton Lakes National Park). Stromatolites do occur throughout the Siyeh/Helena Formation elsewhere in the park, so hikers and backpackers may see a wider range of exposures. The exposures along Going-to-the-Sun Road are marked along the road as Geology Road Stops #2, 7, 8, 13, and 16 (documented in: Raup, Omer B., Robert L. Earhart, James W. Whipple, and Paul E. Carrera, 1983. Geology along Going-to-the-Sun Road, Glacier National Park, Montana. Glacier Natural History Association). On the Goat Lake Trail exposures of stromatolite-bearing limestone occur in the hanging valley in which Goat Lake is situated.

PUBLIC USE: Season and hours: Waterton-Glacier International Peace Park (both Canadian and United States portions) is open year round, but the park visitor season is usually confined to the summer months (mid-June to mid-September or early October). Although weather conditions necessitate closure of some areas, such as Logan Pass, in winter and visitor facilities are limited, the park remains open for winter activities. Access to fossil-bearing exposures is limited to the summer months both by weather (outcrops are covered by snow in winter) and by park seasons of operation; in particular, Logan Pass is open only mid-June through mid-September. **Fees:** Waterton Lakes National Park: $4.00 daily, $9.00/4-day, or $25.00 annual Parks Canada Permit (per vehicle). Glacier National Park: $5.00/vehicle or $2.00/person 7-day pass, or $25.00 annual Golden Eagle Permit (inquire

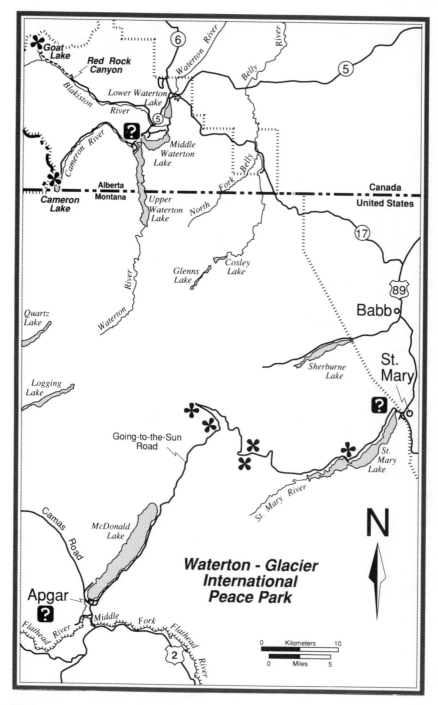

Figure 27. Location of the Precambrian Stromatolite Localities, Waterton-Glacier International Peace Park, Alberta and Montana.

about other park permits). **Food service:** Restaurants and stores are available at park town-sites of Waterton Park, St. Mary, and West Glacier. **Recreational activities:** A full range of park activities and facilities is available including camping (for which an additional fee is charged), backpacking (free permit required), hiking, swimming pool and golf course (Waterton), water activities, and winter activities. Concessioners in the park also provide a variety of activities such as horseback trips, guided backpacking trips, and lake cruises. Because some facilities stay open later in the season depending on the weather, travelers are advised to inquire when making plans. **Restrictions:** Collecting of fossils is prohibited.

EDUCATIONAL FACILITIES: Waterton Lakes National Park: **Visitor Center:** The park information office serves more as a source of information than as a source of interpretation. Interpretive display centers are located at Cameron Lake, Waterton townsite, and Red Rock Canyon. **Visitor Center hours:** The information office is open during the visitor season mid-May to mid-September: daily; from 8:00 A.M. to 6:00 P.M.; with extended hours mid-June to mid-September to 9:00 P.M. Display centers are open June to October: daily; 24 hours. **Fees:** None. **Staff programs:** Interpretive programs are held at park theaters from late June through to Labor Day (daily, 8:30 P.M.) A Junior Naturalist Program is available for children 6–12 years of age.

Glacier National Park: **Visitor Center:** Glacier National Park has visitor centers at Apgar, Logan Pass, and St. Mary. There are ranger stations at several locations including Polebridge, Walton, East Glacier Park, and St. Mary. The visitor centers feature natural history displays which highlight the wildlife and glacial history of the park. **Visitor Center hours:** Visitor centers are open during the visitor season late May (Memorial Day weekend) to September 30 (except that Apgar is open to mid-November): daily; hours vary from center to center but are generally open during normal business hours; with extended hours mid-June to Labor Day usually from 8:00 A.M. to 8:00 P.M. **Fees:** None. **Bookstore:** Bookstores at the visitor centers offer a wide array of books on natural history, especially that of the park, as well as maps and assorted gift items. **Staff programs:** Interpretive programs and naturalist-led hikes are scheduled for the summer visitor season (schedules posted). A Junior Naturalist Program is available for children 6–12 years of age. Native American programs feature speakers from the Blackfeet Tribe. **Note:** The Glacier Institute (P. O. Box 1457, Kalispell, Montana 59903, (406) 752–5222) offers a variety of outdoor seminars for individuals 16 years of age and older (college credit is available for these courses), exploration classes to families with children 10 years of age and older, as well as residential programs for children only.

FOR ADDITIONAL INFORMATION: Contact: Superintendent, Waterton Lakes National Park, Waterton Park, Alberta T0K 2M0, (403) 859–2224 *and* Superintendent, Glacier National Park, West Glacier, Montana 59936, (406) 888–5441. **Read:** (1) Margulis, Lynn. 1988. The ancient microcosm of planet earth. Pp. 83–107 in Osterbrock, Donald E., and Peter H. Raven (editors). Origins and Extinctions. New Haven, Connecticut; London, England: Yale University Press. (2) Raup, Omer B., Robert L. Earhart, James W. Whipple, and Paul E. Carrera. 1983. Geology along Going-to-the-Sun Road, Glacier National Park, Montana. Glacier Natural History Association.

6. Ginkgo Petrified Forest State Park

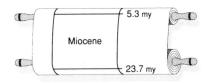

Vantage, Washington

The Laramide Orogeny, the most recent phase of mountain building in the Cordillera, spanned Cenozoic time. Geological processes differed from one part of the Cordillera to another, a collage of events building the western margin of North America that we recognize today. In central and eastern Washington and Oregon, the Columbia Plateau is the major geological feature built during Miocene time. It is defined by the Columbia River Basalts, up to 3000 meters (10,000 feet) thickness of volcanic rocks spread over an area of some 300,000 square kilometers (120,000 square miles) (Figure 28).

The volume of basaltic lava that repeatedly flowed out from deep fissures in the earth was unprecedented in the western Cordillera. Each volcanic episode produced individual layers of lava that are 30 to 150 meters (100 to 500 feet) thick. Some of the layers contain pillow lavas, peculiar pillow-shaped structures that form when lava flows into water. As the molten material comes in contact with water, its surface cools quickly to form a pillow-shaped shell of rock; the lava inside the shell is still much hotter, and it continues to flow. A crack forms in the pillow; the lava flows beyond the pillow and comes in contact with the water; the lava cools quickly, and a new pillow forms. In this way layers of pillows attached end-to-end eventually accumulate.

The Columbia River Basalts preserve an unparalleled diversity of fossilized trees, surprising because the heat of lava—approximately 800 degrees Celsius (1500 degrees Fahrenheit)— destroys organic matter. Yet in central Washington, 15 layers of fossilized wood, 16 to 13 million years old, occur in a 300 meter (1000 foot) thickness of volcanic rock. The apparent contradiction is explained, at least in part, by the occurrence of many of the logs in pillow lavas.

The pillow lavas indicate the presence of standing bodies of water, perhaps formed as previous flows of lava dammed rivers. The fossil logs suggest that fallen trees were carried into the lake, became water-logged, and sank to the lake bottom where they were later entombed in pillows. On the other hand, logs petrified in growth position tell a slightly differ-

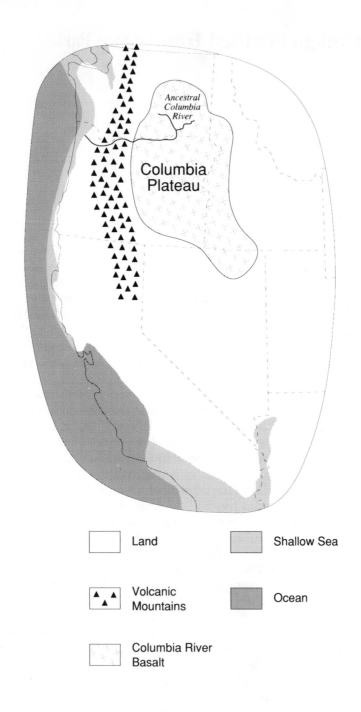

Figure 28. Columbia Plateau during Miocene time.

ent story. It appears that flows of lava dammed a river, formed a lake behind the dam, and flooded large tracts of forests. The flooded trees died, and the lake was floored by a forest of tree stumps. Renewed flows of lava into the lake encased the tree stumps in pillow lavas.

The final step in the fossilization of the logs was their petrification by the silica, which was released into the ground water as the volcanic rocks were weathered. The silica precipitated out of the groundwater into the pores of woody tissue, and the trees were preserved.

Large forests blanketed much of the Pacific coast of North America in Miocene time; a succession of forests, the Russell Forests, reestablished themselves after volcanic eruptions. Several hundred species of trees alone have been identified among the fossil remains. The only known modern analog for such high plant diversity is the tropical rainforest. But the Miocene forests of the Columbia Plateau were not tropical rainforests; they were temperate forests, albeit retaining many of the subtropical plants that had dominated the area in earlier time, developing in response to the decline in temperatures world-wide early in the Miocene. Only abundant rainfall made possible a lush rainforest in warm-temperate conditions. Although the Cascade Mountains had begun to form to the west, there was no appreciable rainshadow yet. Estimates suggest that rainfall probably exceeded 100 centimeters (40 inches) per year, that rainfall was strongly seasonal, most of it falling during the summer.

The Russell Forests grew over a wide area of the Columbia Plateau, and their remains are widely distributed. A comparison with the plants from John Day Fossil Beds National Monument [Site 7] illustrates that climatic conditions changed through time from tropical and subtropical to warm temperate.

A small portion of the Russell Forests is encompassed by Ginkgo Petrified Forest State Park. Many of the trees known from the Russell Forests, among them maple, walnut, elm, sweetgum, horse chestnut, Douglas fir, spruce, and ginkgo, are represented by fossils in the park. Ginkgo is rare in the Russell Forests and only three petrified trees are known in the park, but the park is named for the tree because the first ginkgo to be found in North America was found in this area. The diversity represented by the petrified trees and the unusual fossilization is recognized by the designation of the park as a United States National Natural Landmark.

DIRECTIONS: Access to Ginkgo Petrified Forest State Park is via Interstate 90 east of Ellensburg: from the Vantage Exit (Exit 136) on Interstate 90 follow Vantage Highway (State Route 10) north for 1 kilometer (0.6 mile) to the entrance to the interpretive center and Heritage Area (Figure 29). The fossils are exposed in outcrop in the Natural Area, located 3.4 kilometers (2.1 miles) beyond the entrance to the interpretive center on Vantage Highway.

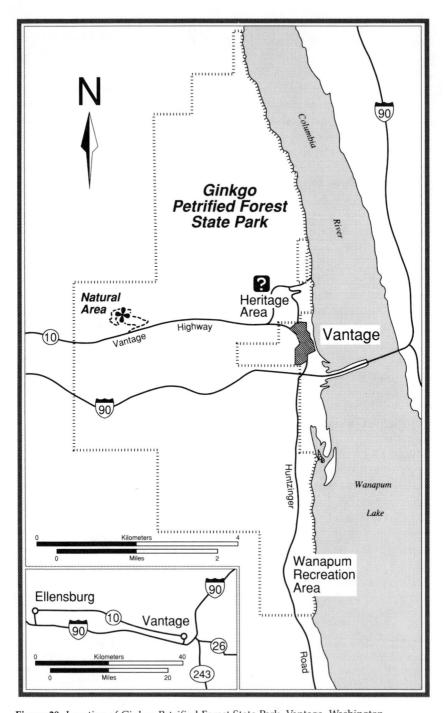

Figure 29. Location of Ginkgo Petrified Forest State Park, Vantage, Washington.

PUBLIC USE: Season and hours: The Ginkgo Petrified Forest State Park is open to the public year round. It is made up of three distinct areas: a Natural Area where the petrified trees are exposed, a Heritage Area where an interpretive center is located, and the Wanapum Recreation Area. **Fees:** None. **Food service:** Restaurants and stores are available in Vantage. **Recreational activities:** Recreational activities are centered south of Interstate 90, on the banks of the Columbia River, at the Wanapum Recreation Area where camping (for which a fee is charged), swimming, fishing, boating, and waterskiing are available. **Restrictions:** Collecting of fossils is prohibited.

EDUCATIONAL FACILITIES: Visitor Center: The interpretive center, located within the Heritage Area, features an extensive exhibit of petrified wood, including polished samples of many of the species found in the park. Various displays and films document the geology of the area. **Visitor Center hours:** Visitor center is open May 16 to September 16: daily; from 10:00 A.M. to 6:00 P.M.; at other times for groups by appointment. **Fees:** None. **Trails:** There are developed trails only in the Natural Area. The self-guiding interpretive trail is short (1.2 kilometers or 0.75 mile) and easy to walk. It leads past logs that have weathered and been excavated out of the flows of lava and are encased in bunker-like cages to protect them from vandalism and further erosion. **Trail hours:** The interpretive trail is open year round: daily; from 8:00 A.M. to dusk, with extended hours April 1 to September 30 from 6:30 A.M.

FOR ADDITIONAL INFORMATION: Contact: Park Ranger, Ginkgo Petrified Forest State Park, Vantage, Washington 98950, (509) 856–2700.

7. John Day Fossil Beds National Monument

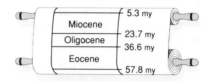

Miocene	5.3 my
Oligocene	23.7 my
	36.6 my
Eocene	
	57.8 my

John Day, Oregon

Cenozoic time—the Age of Mammals! Since the extinction of the dinosaurs 65 million years ago, mammals have been the prominent form of vertebrate life. At John Day Fossil Beds National Monument, an extensive history of the Age of Mammals is represented; the record of successive geological events and changing climatic conditions in consecutive sequences of beds spans the time from 50 to 5 million years ago. Mammals are perhaps the most important fossils recovered in the John Day area, but they are only part of an array that includes complex floras, insects, freshwater and terrestrial mollusks, and freshwater fish. In many cases, the association of fossils approximates the complete biological community as it existed in the geological past.

Fifty million years ago, when the Cenozoic fossil record begins in the John Day area, the western flank of North America was a topographically subdued coastal plain. The highlands of the Cordillera, located to the east, separated the coastal plain from the Great Plains of the interior of the continent. The Cascade Mountains, which are now present to the west, had not yet formed.

By Oligocene time, the western margin of the continent was the focus of renewed tectonic activity. Lava began to erupt, the volcanism an order of magnitude greater than that which built the enormous Cascades volcanoes of the late Cenozoic. Over time, for approximately the next 40 million years, the coastal plain was uplifted and faulted to form a series of highlands and lowlands. Erosion carried sediments from the highlands to be deposited in the lowlands; volcanoes produced an intermittent supply of ash, pumice, and lava. Today we see the remnants of the interbedded sediments and volcanic debris.

Geological events had designed an evolutionary experiment. The biological community of the coastal plain in the John Day area was isolated, separated by physical barriers from the cosmopolitan communities of plants and animals of the Great Plains. Even though temperatures were declining world-wide during the Cenozoic, in the John Day area the cooling was magnified by tectonic uplift; the climate changed over time from tropical to temperate. Simultaneously, as giant volcanoes rose in the west and a rain shadow developed, the amount of precipitation decreased. The biota, geographically isolated and victim to increasingly cooler and drier conditions, evolved in distinctive ways, giving rise to many species unique to the John Day area. It is an exciting result of an evolutionary experiment that paleontologists can track the dynamics of faunal and floral evolution within this closed system over time.

John Day Fossil Beds National Monument comprises three distinct and widely separated units: Clarno, Painted Hills, and Sheep Rock. Each unit represents a portion of the Cenozoic record of central Oregon; each highlights the major characteristics of biological change.

The fossils at Clarno illustrate early Cenozoic time when central Oregon was a warm, moist coastal plain dominated by a tropical or subtropical rainforest of palms, cycads, evergreen oaks, and figs. The 45 million year old Clarno Nut Beds are a famous source of plant remains. The Hancock Mammal Quarry, about 40 million years old, shows it was home to a variety of herbivorous mammals: extinct titanotheres, tiny ancestors of modern horses, tapirs, aquatic and terrestrial rhinoceroses.

With time, the annual mean temperatures dropped, and rainfall declined; a deciduous forest of oaks, bald cypresses, and dawn redwoods replaced the tropical rainforest. At Painted Hills, the leaves of the forest are preserved in layers of volcanic ash. This flora, the Bridge Creek Flora, is approximately 30 million years old. It is noteworthy that tropical species are absent. Fossils of animals from this time are rare, but the most common browsing herbivores of the Oligocene, the oreodonts, are represented. Oreodonts were true artiodactyls, short and stout in appearance, with a complex dental morphology suggesting that they were true ruminants.

Within 10 million years, as the rainshadow expanded, grasslands began to replace the forests, and the vertebrate fauna began to look modern. The transition is recorded at Sheep Rock. Oreodonts continued to be the prominent browsing herbivores, but others are new: early relatives of modern horses, such as *Miohippus*, the three-toed browser with teeth that were becoming more complex and elongated; early carnivores, including dogs and saber-toothed cats; pig-like entelodonts; and many rodents both large and small, including squirrels, pocket gophers, and the bizarre horned species, *Mylagaulodon*. Grasslands came to dominate the John Day area, and grazers became the common mammals.

Each unit at John Day Fossil Beds National Monument is unique. Sheep Rock is the most developed and has the most interpretive information. The outcrops of sedimentary and volcanic rock are deeply eroded, and the scenery is rugged. At Painted Hills, as the name implies, the low hills of weathered volcanic ash are colored various shades of red, pink, gold, tan, and black, colors that vary in hue and intensity with the light and moisture. At Clarno, buff-colored cliffs, exposures of ancient mudslides, dominate.

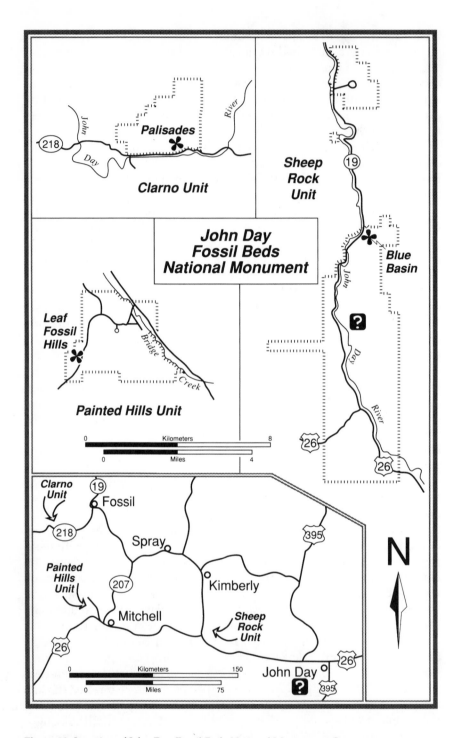

Figure 30. Location of John Day Fossil Beds National Monument, Oregon.

DIRECTIONS: Access to the John Day Fossil Beds National Monument (Sheep Rock Unit and Painted Hills Unit) via United States Route 26 is 61 kilometers (38 miles) west of John Day to the junction with State Route 19 at the Sheep Rock Unit (Figure 30). Follow State Route 19 north 3.4 kilometers (2.1 miles) to Sheep Rock Visitor Center. To reach the Painted Hills Unit continue west on United States Route 26. Access is 5.7 kilometers (3.5 miles) west of Mitchell (junction of United States Route 26 and State Route 207), proceed north from United States Route 26 for 9.8 kilometers (6 miles) to the park boundary. Note that access from the west on United States Route 26 is via Prineville, 77 kilometers (48 miles) to Mitchell. Access to the Clarno Unit via State Route 218 is 32 kilometers (20 miles) west of Fossil or 38 kilometers (23 miles) east of Shaniko (on United States Route 97).

PUBLIC USE: Season and hours: The three units of John Day Fossil Beds National Monument are open to the public year round. Note that portions of the monument as illustrated are private property: the southern quarter of the Sheep Rock Unit and a portion north of the visitor center, the northeastern half of the Clarno Unit, and small portions of the Painted Hills Unit, in particular, the fossil area in the northwest corner. **Fees:** None. **Recreational activities:** Hiking, picnicking, nature activities, and fishing in season with state licence (but not hunting) are available. Camping is not available on site. **Handicapped facilities:** The visitor center and park headquarters are accessible by wheelchair. **Restrictions:** Collecting of fossils is prohibited.

EDUCATIONAL FACILITIES: Visitor Center: John Day Fossil Beds National Monument features a visitor center in addition to the park headquarters in John Day. Each has interpretive displays and exhibits fossils. The visitor center, located at the Sheep Rock Unit, has a fossil preparation laboratory which is designed for public viewing. **Visitor Center hours:** The visitor center is open year round: daily March 1 to November 30 from 8:30 A.M. to 5:00 P.M., with extended hours in summer to 6:00 P.M.; Monday to Friday in winter, December 1 to February 28, from 8:30 A.M. to 5:00 P.M. **Fees:** None. **Bookstore:** Outlets at both the park headquarters and the visitor center offer a limited selection of books on the natural and human history of the John Day area. **Trails:** There is access to fossils (or replicas) *in situ* in each of the three units. In each case a short, self-guiding walk leads to interpretive displays of fossils characteristic of that particular formation. The trails are open year round, 24 hours daily. In the Sheep Rock Unit, the trails (with replicas of representative fossils) are in the Blue Basin area located north of the visitor center; access to the trail head is at the parking area located adjacent to State Route 19. In the Painted Hills Unit, fossils can be seen at Leaf Fossil Hills located in the southwest portion adjacent to the main park road that crosses the unit. In the Clarno Unit, access to the Palisades fossil area is from the picnic area at the park entrance and at the trail head parking area adjacent to State Route 218. **Interpretive sign:** Interpretive panels provide information at Sheep Rock, Clarno, and Painted Hills units.

FOR ADDITIONAL INFORMATION: Contact: Superintendent, John Day Fossil Beds National Monument, 420 West Main Street, John Day, Oregon 97845, (503) 575–0721.

8. Fossil Lake

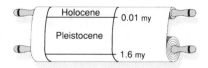

Christmas Valley, Oregon

The last great ice sheet of the Pleistocene Epoch blanketed the northern latitudes of North America between 100,000 and 15,000 years ago, and glaciers were present on many of the high mountain peaks in southern latitudes. All along the ice front as the glaciers melted, water accumulated in glacial lakes. In areas beyond the extent of glacier ice, the climate was less extreme than today, the precipitation higher. As a result, in landlocked areas in the interior of the Cordilleran Mountain Belt, which today are dominated by sand dunes and arid sagebrush, ephemeral glacial lakes developed (Figure 31).

Ephemeral lakes, called playa lakes, form when sudden and sporadic rainfall produces temporary streams flowing toward low areas. Such streams usually carry, in addition to sediment, high concentrations of dissolved minerals. Between periodic injections of stream water, the water in the playa lakes evaporates, salinity increases, and eventually salts can be precipitated. Repeated episodes can produce great thicknesses of salt on salt flats.

Fossil Lake was a persistent playa lake during late Pleistocene time.[17] The region received much more rainfall than it does today, and the lake was a long-standing body of water refreshed by regular rainfall and runoff. The history of Fossil Lake may well be told in the fish, the most common and abundant fossils. They are found in distinct horizons within the sediment, in fish beds that formed when evaporation increased the salinity of the lake water beyond the tolerance of the fish and caused mass deaths.

Fossil Lake is noteworthy among Pleistocene fossil areas for the most diverse and abundant Pleistocene bird fauna in North America, rivalled only by the carnivorous and scavenging birds preserved in the tar pits of Rancho La Brea at Hancock Park [Site 43]. Fossils of birds are uncommon in any geological time: the bones of birds are so delicate that they

---

[17]The Pleistocene Fossil Lake of Oregon is not to be confused with the Eocene Fossil Lake of Wyoming featured at Fossil Butte National Monument [Site 15].

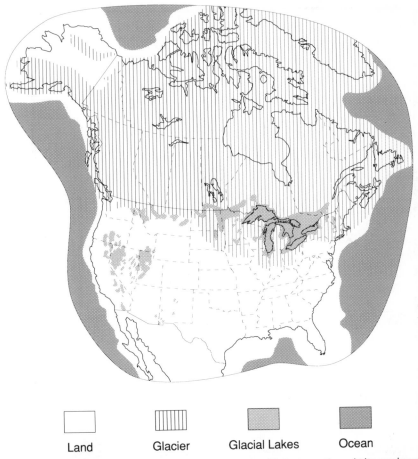

Land	Glacier	Glacial Lakes	Ocean

Figure 31. Paleogeography of North America during late Pleistocene time. At its maximum extent, glacier ice covered most of North America. Glacial lakes were present along the ice front, and playa lakes developed within the highlands of the Cordillera.

are easily destroyed by geological processes long before they can be deposited and fossilized. At Fossil Lake, however, the bones were preserved in organic muds and sands laid down in the quiet and sheltered environments that the ancestral playa lake offered.

The majority of the birds from Fossil Lake are shore birds and water birds. The lake, no doubt, served migratory birds in Pleistocene time as a stop along their north-south flyway. In that case, large numbers of birds would have used the stop, and such use would account for the exceptional fossil record.

Land animals, too, came to the lake, the most reliable source of water in the region. Many died nearby and are preserved in the lake sediment. The bones of mastodons, mammoths, horses, and camels are common.

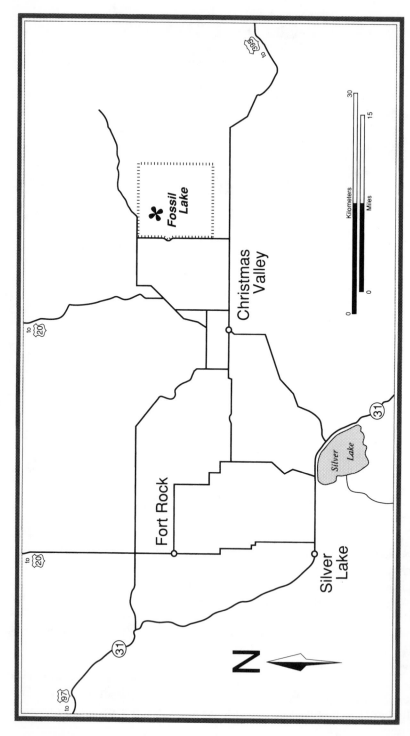

Figure 32. Location of Fossil Lake, Oregon.

A walk among the sand dunes and sagebrush at Fossil Lake is littered with glimpses into the past. Blowouts among the dunes invariably expose bones on the hardpan. The hardpan itself can contain many fish fossils. How startling the impact of climatic change over a geologically short—and humanly comprehensible—period of time! In central Oregon the climate has changed very little over the last 100,000 years except that rainfall has declined appreciably. That modest amount of change, however, has had a dramatic effect on the diversity of animal life.

The Bureau of Land Management administers the land encompassing Fossil Lake. In designating it an Area of Critical Environmental Concern and a Natural Research Area and encouraging renewed research interests, the Bureau of Land Management has recognized the unique aspects of Fossil Lake.

DIRECTIONS: Access to Fossil Lake is via the town of Christmas Valley which is located on a secondary road that extends between State Route 31 and United States Route 395 (junction of State Route 31 and the secondary road is 14.5 kilometers or 9 miles east of Silver Lake, distance to Christmas Valley is 24 kilometers or 15 miles; junction of United States Route 395 and the secondary road is 13 kilometers or 8 miles southwest of Wagontire, distance to Christmas Valley is 77 kilometers or 48 miles). To reach Fossil Lake from Christmas Valley proceed east along the secondary road for 12 kilometers (7.5 miles), then north for 10 kilometers (5 miles) to one entrance or north for 14.2 kilometers (8.8 miles) and then east for 7 kilometers (4.4 miles) to second entrance (Figure 32).

PUBLIC USE: Season and hours: The Fossil Lake area is open year round and access is limited only by weather conditions (such as rain) which make the dirt roads impassable. The Fossil Lake area is fenced and can only be entered on foot. **Fees:** None. **Restrictions:** Collecting of fossils is prohibited.

EDUCATIONAL FACILITIES: Interpretive sign: A sign at each of the two major entrances to Fossil Lake identifies it as a Bureau of Land Management designated area and highlights its paleontological significance. **Trails:** A 1.6 kilometer (1 mile) trail leading into the area from the second entrance is not interpreted.

FOR ADDITIONAL INFORMATION: Contact: Bureau of Land Management, Lakeview Resource Area, Lakeview District Office, P. O. Box 151 (1000 Ninth Street South) Lakeview, Oregon 97630, (503) 947–2177. **Read:** Allison, Ira S. 1966. Fossil Lake, Oregon: Its Geology and Fossil Faunas. Corvallis, Oregon: Oregon State University Press.

9. Hagerman Fossil Beds National Monument

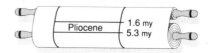

Hagerman, Idaho

In southern Idaho in Pliocene time, lava episodically poured out through fissures in the earth and spread laterally over pre-existing sediments; the volcanic activity of the Laramide Orogeny in the central Cordillera had shifted eastward from Oregon and Washington. The east-west trending Snake River Plateau formed (Figure 33), an extension of the Columbia Plateau that had developed in the Miocene (Ginkgo Petrified Forest State Park [Site 6]).

A thick sequence of fossiliferous sediments interbedded with ash deposits and flows of lava, the Glenns Ferry Formation, is exposed along the banks of the Snake River and in adjacent erosional gullies. At Hagerman, the outcrop is 170 meters (550 feet) thick and extends laterally for more than 8 kilometers (5 miles). The sediments of the Glenns Ferry Formation were deposited in lakes on floodplains and in stream channels; but repeatedly, volcanic ash rained down on the sediments, and flows of lava interrupted deposition. The latter are caps of rock that protect underlying sediments from erosion and are horizontal marker beds that can be dated radiometrically. They are the basis for a detailed chronology spanning half a million years (3.7 to 3.2 million years old) for the fossils in the area.

Global temperatures cooled substantially during late Tertiary time, evidenced by the appearance of glaciers over the South Pole as early as Miocene time. As a direct result of the decline in temperature and the extensive rainshadow of the Cordilleran Mountain Belt, temperate and cool-temperate grasslands and savannahs spread over much of the western and interior portions of North America. These were rich biological zones capable of supporting large populations of animals.

Hagerman Fossil Beds National Monument contains hundreds of fossil sites, and the fauna is the most abundant and diverse of its age in the world. The vertebrate component of the fauna illustrates the point: more than 100 species of vertebrates are known at Hagerman, and more than half of those are mammals; the wealth of the mammals is underscored by the fact that, of all the fossil mammals of that age (± 3.5 million

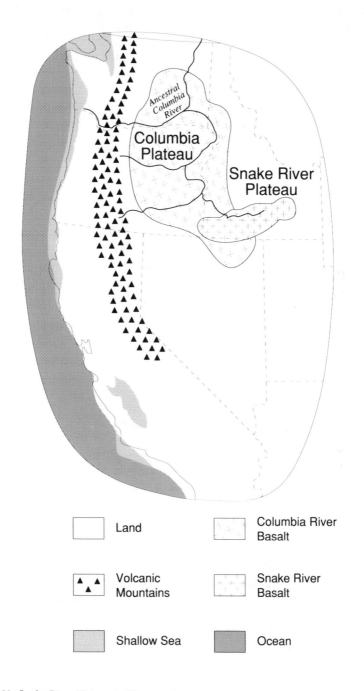

Figure 33. Snake River Plateau in Pliocene time.

119

years old) known in North America, approximately half are found at Hagerman. Plants are found in some exposures of the Glenns Ferry Formation, in others freshwater mollusks are abundant (Sand Point Fossil Area [Site 10]). Hagerman Fossil Beds National Monument, however, is famous world-wide for the fossil Hagerman horse, *Equus simplicidens*, the state fossil of Idaho. From a single quarry, the Horse Quarry, more than 150 nearly complete skulls of *Equus simplicidens* have been collected.

Equus simplicidens was adapted to life on the grasslands. It had elongate limbs for speed; all the weight was born by the central digit of the fore and hind limbs, the lateral digits lost. It had teeth designed as complex grinders to process large quantities of abrasive food; they were elongate and ever-growing, assuring that in spite of continuous wear they would last a lifetime. Small in stature, it was very much like a modern zebra.

Many grasslands animals are found with the Hagerman horse. Among them are deer, antelope, and camels. There are also many rodents: squirrels, pocket gophers, kangaroo mice and rats, voles, lemmings, mice, beavers; as a group, the ecologically most important grasslands animals. Carnivores, too, were plentiful given such an abundance of prey.

Camels were native to North America. They evolved here early in the Tertiary (Eocene-Oligocene) and became important components of the Tertiary faunas of North America; for example, there are various species of camels found at Hagerman. Only in the Pliocene did certain camels expand their ranges across the Bering region into Asia and in the Pleistocene across Panama into South America when land bridges were established between the continents. The irony is that camels became extinct in their homeland some 10,000 years ago.

The visible aspects of the sediments exposed along the Snake River valley belie their paleontological importance. Fossils are not usually exposed unless spring runoff, occasional heavy rainfalls, or strong winds have eroded the sediments. Nonetheless, the potential for research and public interpretation here is tremendous because the fossils are so diverse and abundant, the thick sequence of sediment is fossiliferous throughout, good marker beds are present at various intervals, and the volcanic beds can be accurately dated.

Hagerman Fossil Beds National Monument is a new addition to the National Park Service (as of November, 1989). Development for public access and use is only just beginning, but plans for the next five years include the building of a visitor center and the development and implementation of various interpretive programs. Visitors are welcome in the meantime.

DIRECTIONS: The Hagerman Fossil Beds National Monument is located on the west bank of the Snake River, directly across the river from the town of Hagerman, and it extends southward to the Dolman Rapids (Figure 34). There is no formal access road into the site

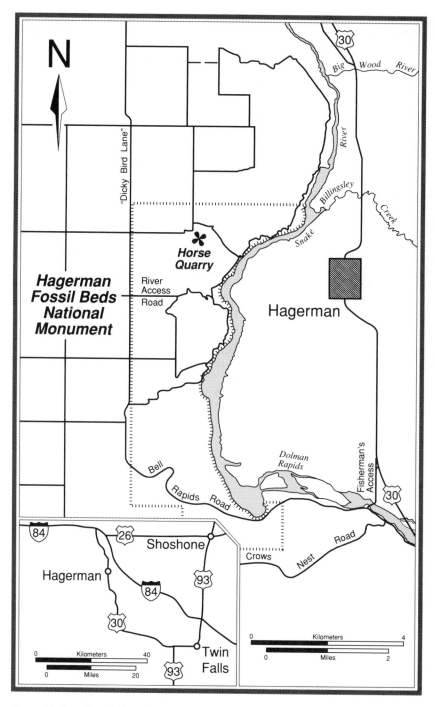

Figure 34. Location of Hagerman Fossil Beds National Monument, Idaho.

as yet, and the locations of various facilities are yet undetermined. The most convenient access to the west side of the Snake River is along United States Route 30 south of Hagerman 4.7 kilometers (2.9 miles) to Fisherman's Access Road (Old Highway 30) then 1 kilometer (0.6 mile) to Owsley Bridge. Once across the bridge, proceed west and north along Bell Rapids Road for 8.7 kilometers (5.4 miles). Bell Rapids Road proceeds west from this point. Continue north along "Dicky Bird Lane" 3.2 kilometers (2 miles) to the river access road and another 1.6 kilometers (1 mile) to the Horse Quarry access. Each access road leads due east. Beyond the access roads, exploration is by foot.

PUBLIC USE: Season and hours: Hagerman Fossil Beds National Monument is a new national monument. The area remains open to the public year round, but access is limited by snow in mid-winter. Until the monument is developed for public use the most effective way to see the area and learn about paleontology is to participate in a tour to the site. **Fees:** None. **Restrictions:** Collecting of fossils is prohibited.

EDUCATIONAL FACILITIES: Museum: Hagerman Valley Historical Museum is the best interim source of information in Hagerman. It features a mounted replica of the Hagerman horse, *Equus simplicidens*, and the plaque commemorating the Hagerman Fossil Area as a National Natural Landmark is on display. The museum is cooperating with the National Park Service to sponsor tours to the fossil area. **Fees:** None. **Special group activities:** Public tours for groups can be arranged by contacting the office of Mr. David Pugh, Superintendent, Hagerman Fossil Beds National Monument, but advance notice is required. At present, the office is located in Twin Falls, Idaho; an on-site office is being planned.

FOR ADDITIONAL INFORMATION: Contact: Superintendent, Hagerman Fossil Beds National Monument, 963 Blue Lakes Boulevard, Twin Falls, Idaho 83301, (208) 733–8398. **Read:** (1) Bonneckson, Bill, and Rory M. Breckenridge (editors). 1982. Cenozoic Geology of Idaho. Idaho Bureau of Mines and Geology, Moscow, Bulletin 26. (2) Malley, Terry. 1987. Exploring Idaho Geology. Boise, Idaho: Mineral Land Publications.

10. Sand Point Fossil Area

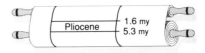

Pliocene 1.6 my 5.3 my

Hammett, Idaho

The Glenns Ferry Formation is a thick sequence of floodplain and stream channel sediments interbedded with volcanic ash deposits and lava flows. It is exposed along the banks and in the erosional gullies of the Snake River. Ninety meters (300 feet) of sediments are exposed at the Sand Point Fossil Area, and mollusks are readily visible in layers within the bedded sediments.

The Glenns Ferry Formation is precisely dated because it contains layers of volcanic ash and lava. At Sand Point Fossil Area, the Glenns Ferry Formation and the shell beds are about 3.1 million years old, slightly younger than the sediments of the Glenns Ferry Formation exposed at Hagerman Fossil Beds National Monument [Site 9].

The Bureau of Land Management, which administers the Sand Point Fossil Area, has recognized the importance of the area and has designated it a Bureau of Land Management Area of Critical Environmental Concern. The fossils here, as at Hagerman, are important for their research potential and are a focus of public interest; in contrast to the vertebrate fossils at Hagerman, however, the fossils at Sand Point are readily accessible and well exposed, abundant and well preserved. Because they are so accurately dated, the mollusks are a vehicle for studying the processes of evolutionary change over relatively short periods of geological time. The Sand Point snail may not be as exciting as the Hagerman horse, but the snails serve to complete a picture of life in the past, life that was not confined exclusively to vertebrates.

DIRECTIONS: The Sand Point Fossil Area is located on the south bank of the Snake River south of the town of Hammett (Figure 35). Access is by way of State Route 78 south of Hammett 6 kilometers (3.7 miles) and across the Snake River at Indian Cove Bridge. Hodco Road leads from State Route 78 eastward along the river into the Sand Point Fossil Area. Follow Hodco Road 3.4 kilometers (2.1 miles) until it forks south of Sparling Island (Hodco Road leads up the hill from this point). Sailor Creek Access Road continues along the river to Sailor Creek, and beyond the creek the access road (part of the Historic Oregon Trail) leads past the fossil sites and then climbs up out of the valley. The fossil sites are 4 kilometers (2.5 miles) beyond the fork in Hodco Road.

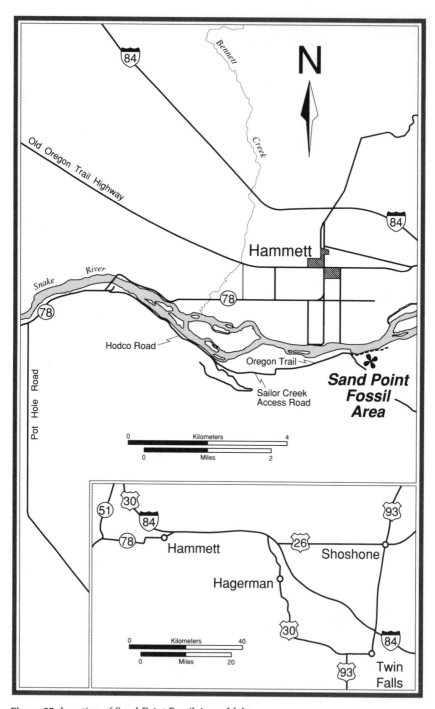

Figure 35. Location of Sand Point Fossil Area, Idaho.

PUBLIC USE: Season and hours: The Sand Point Fossil Area is on public lands administered by the Bureau of Land Management. It is open to the public year round. **Fees:** None.

FOR ADDITIONAL INFORMATION: Contact: Bureau of Land Management, Boise District Office, 3948 Development Road, Boise, Idaho 83705, (208) 334-1582.

11. Malm Gulch Fossil Wood Area

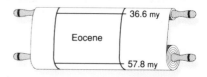

Challis, Idaho

The Malm Gulch petrified forest in central Idaho is preserved in the volcanic rocks and ash of the Challis Volcanics, rocks that extend over several thousand square kilometers of central Idaho. In the Challis area, the volcanics are mainly tuffs and ash, but they interfinger laterally with lava rocks.

Petrified trees are the most visible fossils found at Malm Gulch. Many are preserved in growth position, but some have now tumbled over as erosion has exposed them. Within a thickness of 55 meters (180 feet) of tuffs and ash are six horizons of petrified trees. Some of the trees are up to 3 meters (10 feet) in diameter, a size indicating that the relative length of time between successive ash falls was sufficient for mature trees to become established. Leaves, stems, and fruits can be found with insects, freshwater invertebrates, and fish preserved in ash that settled in the bottom of small lakes.

The Challis Volcanics and the fossils they contain are Eocene in age, approximately contemporaneous with volcanic rocks of the Clarno Unit of John Day Fossil Beds National Monument [Site 7] and with the Petrified Forests in Yellowstone National Park [Site 13]. The forests of deciduous hardwood trees, dawn redwoods, and pines that grew in the Challis area approximately 40 million years ago were significantly different, however, from the tropical forests dominating the landscape along the coastal plain to the west. There was less rainfall in the interior, the elevation was higher and the temperatures were cooler. Nonetheless, the climatic conditions were much more equable than at present; in the Eocene, the region was subtropical or warm-temperate and moist, and the vegetation lush.

DIRECTIONS: The Malm Gulch Fossil Wood Area is located near Challis, Idaho (Figure 36). Access is presently limited to foot traffic. Visitors are requested to contact the Salmon District Office of the Bureau of Land Management for current information and regulations about the site and directions to it.

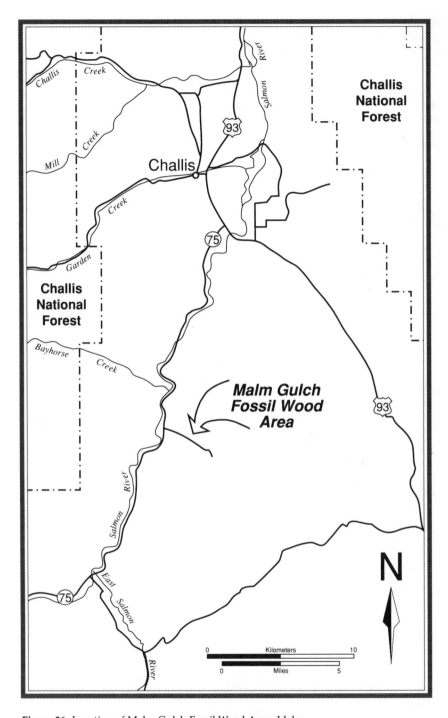

Figure 36. Location of Malm Gulch Fossil Wood Area, Idaho.

PUBLIC USE: Season and hours: The Malm Gulch Fossil Wood Area is on public lands administered by the Bureau of Land Management. It is open year round, but winter conditions make it inaccessible from November through March or April. **Fees:** None. **Restrictions:** Collecting of fossils is prohibited.

FOR ADDITIONAL INFORMATION: Contact: Bureau of Land Management, Challis Area Manager, Salmon District Office, P. O. Box 430, Salmon, Idaho 83467, (208) 756–2201. **Read:** (1) Bonneckson, Bill, and Rory M. Breckenridge (editors). 1982. Cenozoic Geology of Idaho. Idaho Bureau of Mines and Geology, Moscow, Bulletin 26. (2) Malley, Terry. 1987. Exploring Idaho Geology. Boise, Idaho: Mineral Land Publications.

12. Dinosaur Nest Sites of the Willow Creek Anticline

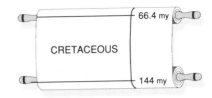

66.4 my

CRETACEOUS

144 my

Choteau, Montana

Some dinosaurs were very mammal-like: many species were gregarious, social creatures; many were agile and swift, others were migratory with the seasons. The fossil evidence is incontrovertible. A critical piece of evidence documenting the behavior of dinosaurs is found in the Willow Creek Anticline near Choteau, Montana.

In Late Cretaceous time when the Western Interior Seaway occupied most of the western interior of North America, large herds of dinosaurs roamed along its western shore. Sediment-laden rivers flowed from the rising mountains to the west, across broad floodplains, and into the seaway; swamps and deltas formed near the shore. Such deposits are rich with dinosaur fossils. In drier, upland areas where erosion is more common than deposition, the deposits are thinner, and fossils are fewer.

The Two Medicine Formation of the Willow Creek Anticline was deposited in an arid environment well above sea level. It is made up of variegated red, purple, and brown sediments, which were deposited in streams and lakes, and contains abundant volcanic ash. It is contemporaneous with the Judith River Formation in Alberta (Dinosaur Provincial Park [Site 2]). The environment in which each formation was deposited and the nature of the fossils each contains reveal the complexity of the Cretaceous landscape: arid, upland valleys coexisted with broad river-swamp-delta complexes. The area that is now western Montana, 75 to 70 million years ago, was a broad valley between the rising mountains to the west and a range of highlands to the east. The shoreline of the Western Interior Seaway, some 350 kilometers (200 miles) distant, lay to the east beyond the highlands (Figure 21). The mountains rising in the west had created a rainshadow, and within the valley, rainfall was seasonal and episodic.

Sediments from the highlands on either side of the valley were carried into the valley by ephemeral streams. The lakes were shallow and subject to periodic drying. Nesting sites and nests, which occur in distinct horizons interbedded by layers of volcanic ash, are the most significant fossils in the Two Medicine Formation. Both eggshells and bones are abundant. Within some nests are eggs; within some eggs, the tiny bones of developing embryos. In other nests are the bones of juvenile dinosaurs in various stages of early growth; still others have both eggs and juveniles. Thus among the egg shells and volcanic ash is the evidence that at least some dinosaurs migrated to this higher and drier ground, away from the swamps and floodplains near the seaway, to nest and raise their young. Three species, two herbivores and a carnivore, are well represented in the Two Medicine Formation; apparently the only ones to use the area for nesting.

The best known dinosaur from the Willow Creek Anticline is the hadrosaur *Maiasaura*, the state fossil of Montana. Its nests are concentrated in large rookeries to which an entire herd would return regularly and leave only when the juveniles were old enough to travel. The sediments that enclose the *Maiasaura* rookeries are distinctly green in color, harder and more dense than surrounding rocks. The contrast is due to different iron minerals in the soil, chemical changes brought about by the high concentration of vegetable and fecal material in and around the nests. Clearly this duck-billed dinosaur nurtured its young in the nest in true bird-like fashion, foraging for food and returning to the nest to regurgitate partly digested plants for its hatchlings.

A small bipedal herbivore, a member of the hypsolophodontid family (Suborder Ornithopoda), is also well represented in the Two Medicine Formation. The untrampled, unsoiled condition of the nests reveals that the young were able to fend for themselves immediately upon hatching, that they left the nest to forage independently.

Nearby are the nest sites of a small carnivorous dinosaur called *Troodon*. The carnivores, too, apparently nested in colonies and cared for their young. Certainly the *Maiasaura* colonies nearby and the precocious hypsolophodontids were a ready source of food.

All of the fossil sites in the Willow Creek Anticline are on land that is privately owned. The site of the original discovery, a knoll called Egg Mountain, is now owned by The Nature Conservancy and is part of the Pine Butte Swamp Preserve. Egg Mountain is being systematically excavated, and a wealth of data and fossils is being collected. Other dinosaur sites are on private ranch land.

DIRECTIONS: Choteau, located at the junction of United States routes 89 and 287, is 85 kilometers (50 miles) northwest of Great Falls (Figure 37). All the fossil sites are within 20 kilometers (12 miles) of Choteau. For access and further information contact the Pine Butte Swamp Preserve or the Old Trail Museum.

PUBLIC USE: Season and hours: All fossil sites in the area are on land that is privately owned, and direct public access is restricted. During the summer, both the Pine Butte Swamp Preserve and the Old Trail Museum run public tours of fossil sites. Regular tours are scheduled daily (except Saturday for the Old Trail Museum only) or can be arranged by appointment. **Fees:** Tours of the Egg Mountain locality on the Pine Butte Swamp Preserve are free. The Old Trail Museum tours a number of sites for which it charges $7.50 for

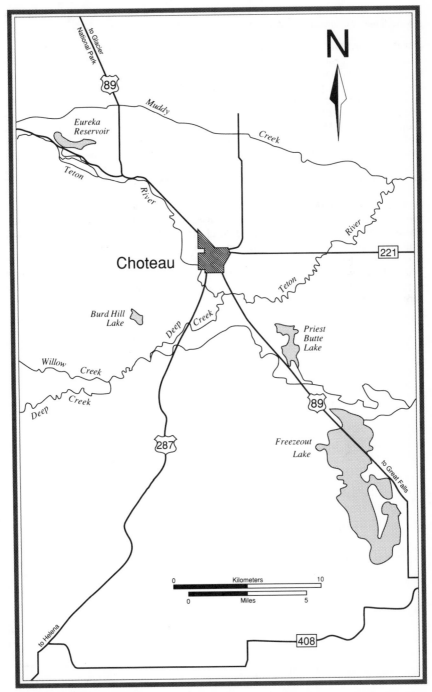

Figure 37. Location of Choteau, Montana, for access to the Dinosaur Nest Sites of the Willow Creek Anticline.

adults and $4.00 for children under 12 years of age. **Food service:** Restaurants and stores are available in Choteau. **Restrictions:** Collecting of fossils is prohibited.

EDUCATIONAL FACILITIES: Museum: Old Trail Museum is a small museum that features an assortment of rocks and fossils, most of which have been collected locally. **Museum hours:** The museum is open Memorial Day to Labor Day: daily; from 10:00 A.M. to 6:00 P.M. **Fees:** None. **Special group activities:** The Pine Butte Swamp Preserve offers a summer naturalist program to paying guests at The Nature Conservancy's Pine Butte Guest Ranch, which includes visits to fossil sites being excavated and weekly slide shows given by paleontologists from the Museum of the Rockies. Further information on these programs is available from the Naturalist, The Nature Conservancy, HC 58, Box 27, Choteau, Montana 59422. **Note:** In late summer, a week-long dinosaur dig workshop is held in conjunction with a field course in paleontology sponsored by the Museum of the Rockies, Bozeman, Montana. Enrollment in the workshop is limited so early enquiries are recommended.

FOR ADDITIONAL INFORMATION: Contact: Pine Butte Swamp Preserve, HC 58, Box 34B, Choteau, Montana 59422, (406) 466–5526 *and* Old Trail Museum, Inc., P. O. Box 919, Choteau, Montana 59422, (406) 466-5332. **Read:** (1) Hirsch, Karl F., and Betty Quinn. 1990. Eggs and eggshell fragments from the Upper Cretaceous Two Medicine Formation of Montana. Journal of Vertebrate Paleontology, volume 10, number 4, pp. 491–511. (2) Horner, John R. 1984. The nesting behavior of dinosaurs. Scientific American, volume 250, number 4, pp. 130–137. (3) Horner, John R., and James Gorman. 1988. Digging Dinosaurs. New York, New York: Workman Publishing.

13. Petrified Forests
Yellowstone National Park

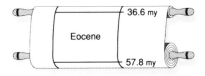

Mammoth Hot Springs Junction, Wyoming

Yellowstone National Park has been shaped by the Laramide Orogeny, the latest major pulse of mountain building in the Cordillera. During Eocene time, intense volcanic activity was the dominant geological process. The petrified forests and diverse other plant remains that are preserved within Yellowstone National Park provide the information necessary to reconstruct the environments and events of that volcanic time. The petrified forests occur in multiple horizontal layers and are buried in mud, ash, and rock debris; leaves, fruits, seeds, and cones are found nearby, preserved in ponds and lakes as volcanic ash slowly settled out of the water.

Northwestern Wyoming in Eocene time was a low, hilly plain crossed by broad river valleys and dotted by lakes. It lay to the east of the highlands of the Cordillera, along the southeastern margin of a broad volcanic belt. The climate was moist and subtropical; deciduous hardwood forests were widespread, as evidenced by the petrified trees in the Challis Volcanics in northern Idaho at Malm Gulch Fossil Wood Area [Site 11]. The quiescence was repeatedly disrupted by volcanic activity that was to form the northern Gallatin Range and the Absaroka and Washburn ranges.

Volcanic explosions produced a rain of ash and rock debris. Molten lava poured out of the volcanoes and solidified into layers. Snow and ice on the upper slopes of the volcanoes melted with the heat of the eruptions and induced landslides and mudflows. Some streams became choked and ponded; others were diverted. Mudflows mowed down forests and carried the trees and stumps downslope in the debris. Volcanic ash settled out of the atmosphere, choked lakes and rivers, and buried plant remains. Very quickly after each eruption, plant growth began again, and forests became reestablished. The process of repeated destruction and rejuvenation produced a sequence of volcanic and sedimentary rocks known as the Lamar River Formation, part of the Absaroka Volcanics.

Out of the destruction of forests came remarkable preservation of fossils. Volcanic debris introduced large quantities of silica into the groundwater, dissolved silica which then readily precipitated from the groundwater into the woody tissue of buried trees and petrified them. The fine volcanic ash settling in lakes and ponds preserved delicate leaves, flowers, fruits, and even insects in layers of mud and ash on the bottom.

Almost 200 species of plants have been identified among the Eocene fossils of Yellowstone National Park. Most are hardwood trees and flowering plants, trees such as sycamores, walnuts, chestnuts, and oaks, a flora similar to that found today in warm-temperate and subtropical southeastern United States. There are also many conifers, primarily spruce and fir, and ferns, all of which are characteristic of much colder climates; and some tropical plants, such as figs and laurels. In addition, some exotic trees are known, extinct relatives of katsura and breadfruit trees, for example. The composition is anomalous; cold-adapted, subtropical or warm-temperate, and tropical species appear to have lived together.

The inconsistency in floral composition is resolved if the present occurrence of fossils is interpreted as representing, not life conditions, but the results of transportation and deposition. The most common plants, the warm-temperate and subtropical hardwoods, must have occupied the warm, low-elevation valleys. The tropical plants were marginal components, remnants from an earlier and warmer geological time; they must have occurred in pockets in the hardwood forest. The cold-adapted spruce and fir would have grown high on the mountain slopes where the elevation was responsible for cool- and cold-temperate conditions. Ashflows and mudflows induced by volcanic eruptions carried the trees from the high mountain slopes down to be deposited among the subtropical plants of the valley.

The petrified forests, layer upon horizontal layer in vertical succession, record the destructive powers of volcanic activity, the regenerative capacity of living organisms, and the tremendous potential for fossilization when circumstances are conjoined. Some trees are preserved in growth position, others are preserved in the muds that transported them; they have been turned to stone by the precipitation of silica in the woody cells. They litter the landscape because they are harder and more resistant to erosion than the surrounding rock. Fossil forests have one advantage that living forests do not—they can withstand the ravages of forest fires such as those that swept much of Yellowstone National Park in 1988.

DIRECTIONS: Yellowstone National Park is located in extreme northwestern Wyoming, overlapping into Montana and Idaho (Figure 38). Access into the park from the south is via United States Route 287/191/89 from Grand Teton National Park; from the east via United States Route 20/16/14 from Cody; from the northeast via United States Route 212 from Cooke City, Montana; from the north via United States Route 89 from Gardiner, Montana;

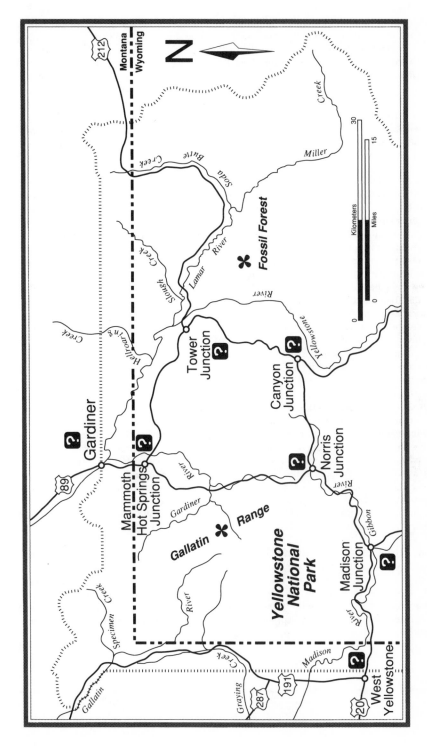

Figure 38. Location of the Petrified Forests in Yellowstone National Park, Wyoming.

and from the west via United States Route 287/191/89/20 at West Yellowstone, Montana. Petrified trees can be seen in two main areas in Yellowstone National Park. In the vicinity of Tower Junction, fossil trees are found on either side of the Yellowstone River and along the Lamar River (Specimen Ridge and Fossil Forest). They also occur in the Gallatin Range (Fossil Ridge) in the northwestern corner of the park. Fossil Ridge can be reached by hiking into the Gallatin Range from Mammoth Hot Springs; Specimen Ridge and Fossil Forest are accessible from Tower Junction and the northeast entrance road, a continuation of United States Route 212. (Note that highway numbers are not assigned or marked on park roads). Some fossil trees are present near roadways (for example, along the road from Gardiner to Cooke City just west of Tower Junction). Details and maps are available at the visitor centers for those visitors who wish to hike into the principal fossil sites.

PUBLIC USE: Season and hours: Yellowstone National Park is open to the public year round, but activity is restricted in the winter. Park roads are open only in summer (about May 1 to October 31 depending on weather conditions) except that the road between Gardiner and Cooke City, Montana is open year round. Most park facilities are open from May 30 to October 15 (but may be limited after September 1). Access to the petrified forests is limited to the summer months by weather conditions (outcrops are covered by snow in winter). High water in the Lamar River may also limit access. Details about trail conditions and closures are available from park rangers. **Fees:** $10.00/private vehicle or $5.00/person seven day pass (valid in both Yellowstone National Park and Grand Teton National Park), or $25.00 annual Golden Eagle Permit (inquire about other park or area permits). **Food service:** Restaurants, stores, and concessions are available in the park. **Recreational activities:** A wide range of park activities is available including back-country hiking (free permit required), horseback riding, boating, fishing (permit required), and camping (for which an additional fee is charged); and winter activities such as skiing and snowmobiling. Private concessioners also provide a variety of activities. **Restrictions:** Collecting of fossils is prohibited.

EDUCATIONAL FACILITIES: Visitor Center: Yellowstone National Park has eight visitor centers/ranger stations with natural history displays, museum and field exhibits. **Visitor Center hours:** The visitor centers/ranger stations are open during the visitor season: daily; hours are variable according to demand; with extended hours Memorial Day to Labor Day. **Fees:** None. **Bookstore:** A wide variety of natural history items including books and maps are available at bookstores in the visitor centers. **Staff programs:** From mid-June through Labor Day park ranger-naturalists offer a wide variety of guided hikes, interpretive programs, and demonstrations. Information on these programs is posted and readily available throughout the park. **Note:** The Yellowstone Institute offers field courses, with university credit available for some, including one called "Fire, Ice and Fossil Forests."

FOR ADDITIONAL INFORMATION: Contact: Superintendent, Yellowstone National Park, P. O. Box 168, Yellowstone National Park, Wyoming 82190, (307) 344–7381. **Read:** (1) Dorf, Erling. 1964. The petrified forests of Yellowstone Park. Scientific American, volume 210, number 4, pp. 106–114. (2) Fritz, W. J. 1980. Reinterpretation of the depositional environment of the Yellowstone "fossil forests." Geology, volume 8, pp. 309–313. (3) Wolfe, Jack. 1978. A paleobotanical interpretation of Tertiary climates in the northern hemisphere. American Scientist, volume 66, pp. 697–703.

14. Minnetonka Cave
Caribou National Forest

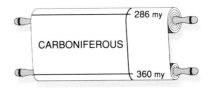

CARBONIFEROUS

286 my

360 my

Saint Charles, Idaho

All along the eastern flank of the Cordillera, from Northwest Territories to Arizona, limestone rocks hundreds of meters thick are exposed. The names vary: in the mountains where limestones form steep, massive cliffs they are usually called the Mission Canyon Limestone (in Canada) or Madison Limestone; in the Grand Canyon [Site 33] where they are flat-lying and exposed at depth they are called the Redwall Limestone. In some places, the limestones have been altered to dolomite. These thick, fossil-rich rocks document a time approximately 350 million years ago, when shallow epicontinental seas were widespread across most of western North America; the lateral distribution of the rocks indicates the minimum extent of the sea (Figure 39). The continent was smaller then, the mountains of the Cordillera had not yet been built, and there was no impediment to the waters of the ancient Pacific as the ocean flooded over the continent. The conditions were ideal for an abundant marine invertebrate fauna: shallow, clear epicontinental seas and a warm, sub-tropical climate.

Thirty million years earlier in Late Devonian time, marine invertebrates, in particular the tabulate-stromatoporoid reef community, had been decimated by mass extinction; but in the seas of Early Carboniferous time, a rich, new invertebrate fauna evolved rapidly. No reef-building organisms evolved; instead, brachiopods, bryozoans, and calcareous algae built low banks and shallow mounds on the sea floor.

The crinoids became so widespread and greatly diversified that Early Carboniferous time can be called the age of crinoids. Ancient relatives of modern sea lilies, crinoids were suspension feeders attached to the sea floor. By means of a long stem, the calyx was held high in the water and feathery arms were used to gather food particles that floated by. So extensive were the crinoids that they formed vast meadows extending for hundreds of square kilometers across the expanse of epicontinental sea.

The growth form of crinoids is responsible for the unique crinoidal limestone. Crinoid stems are made up of stacked plates that resemble a

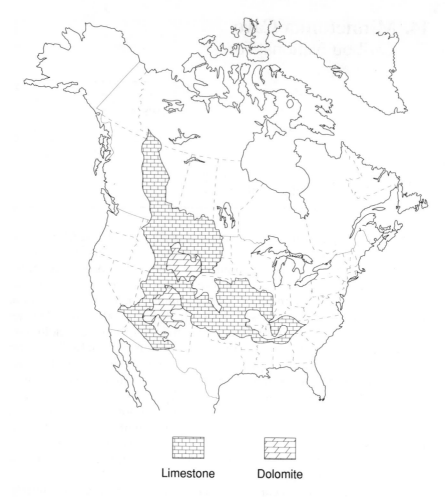

Limestone   Dolomite

Figure 39. Distribution of limestone rocks of Early Carboniferous age in the interior of North America. In some places the limestone has been altered to dolomite.

stack of tiny doughnuts. When the animals died, the plates fell apart and were broken and sorted by the waves. Rare are complete specimens of crinoids, but segments of stems and isolated plates are the dominant constituents of the limestone deposit in many exposures.

Millions of years after the crinoid meadows had vanished, their remains buried by younger layers of sediments, mountains began to form. The layers of sediments were faulted and uplifted; the softer rocks were eroded away, massive cliffs and mountain sides of resistant limestone remain. In Minnetonka Cave, high in the Bear River Range in southeastern Idaho, the limestone is rich with marine invertebrate fossils: crinoids, horn corals, and brachiopods. The fossils are clearly visible in

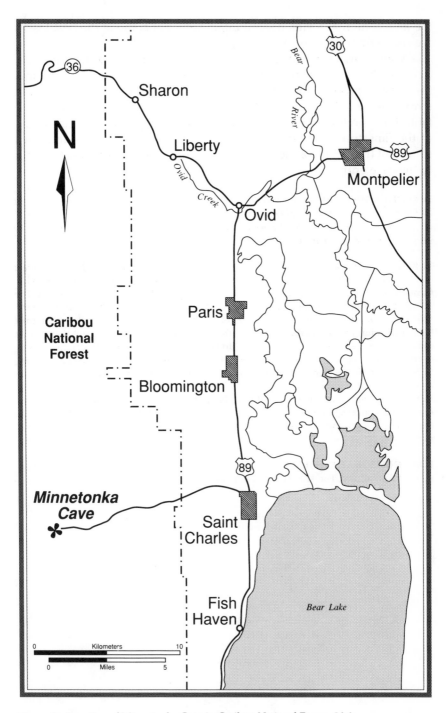

Figure 40. Location of Minnetonka Cave in Caribou National Forest, Idaho.

the limestone that forms the mountain side, but some of the best fossils are seen within the cave itself.

Minnetonka Cave formed along the line of a fault, the Temple Fault, a natural break in the rocks that developed during mountain building. Much later, probably within the last two million years, an underground stream flowed along the fault line and dissolved the limestone along the fault. But it preferentially dissolved the lime muds that formed the matrix of the limestone and left the fossils in relief. Crinoid stems and horn corals dot the ceiling of the cave. Outside, brachiopods litter the footpath to the cave entrance.

DIRECTIONS: Follow United States Route 89 south from Montpelier toward Bear Lake for 29 kilometers (18 miles) to Saint Charles. Minnetonka Cave is situated at the end of a 16 kilometer (10 mile) paved Forest Service road leading west from United States Route 89 into Caribou National Forest (Figure 40).

PUBLIC USE: Season and hours: Minnetonka Cave is located in Caribou National Forest and is administered by the United States Department of Agriculture, Forest Service. Caribou National Forest is open to the public year round, but winter snowfall is heavy. The access is best from about mid-April to mid-November. The United States Forest Service, Montpelier District operates regular tours of Minnetonka Cave. Fossils are not a formal part of the tour, but the staff are aware of the fossils and are happy to point them out in the cave. Tours run early June (date varies) to Labor Day: daily; from 8:30 A.M. to 5:30 P.M. every half hour. **Fees:** Caribou National Forest: none; Minnetonka Cave tours: $2.50 adults, $1.50 children. **Recreational activities:** Caribou National Forest offers camping (for which a fee is charged) within a short distance of Minnetonka Cave; water activities, including fishing, and winter activities are featured. Note, however, that Caribou National Forest rangers are concerned about adverse affects of visitors in winter on big-game animals in their wintering grounds. Visitors are advised to check with the Forest Service when planning winter activities.

FOR ADDITIONAL INFORMATION: Contact: Montpelier District, United States Department of Agriculture, Forest Service, Montpelier, Idaho 83254, (208) 847–0375.

15. Fossil Butte National Monument

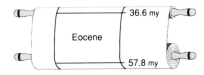

Kemmerer, Wyoming

Fossil Butte National Monument highlights characteristic geological processes active during the Laramide Orogeny in the Central Rockies, specifically southern Wyoming, eastern Utah, and western Colorado. Tectonic uplift began late in Cretaceous time and continued for about 20 million years, uplifting large blocks of basement rock to form mountain ranges, the blocks between them gradually subsiding to form inter-montane basins. Sediment from the rising highlands began to accumulate in the basins. These events were distinct from the volcanism that characterized contemporaneous mountain building elsewhere in the Cordillera (for example, volcanism in the Absaroka and Gallatin ranges of Yellowstone National Park [Site 13]).

Intermontane basins are widely distributed in the Central Rockies: the Green River Basin of southwestern Wyoming, the Powder River Basin of northern Wyoming, the Uinta Basin of eastern Utah, and the San Juan Basin of northern New Mexico (Figure 41). They formed early in the Eocene; they were large and were occupied by lakes throughout most of Eocene time. The sediments they contain are richly fossiliferous.

The Green River Basin is typical. Early in Eocene time, Lake Gosiute occupied a small area in the center of the Green River Basin; sedimentation throughout the basin consisted of fluvial and alluvial conglomerates, sandstones, and siltstones. The Wasatch Formation, varying in thickness from 800 to 1800 meters (2600 to 6000 feet), is characterized by vibrant hues of red and orange; the sands and coarse sediments that accumulated on an alluvial plain are interbedded with fine-grained, green and buff river floodplain deposits. The rocks yield a distinctive assemblage of terrestrial mammals including primitive primates; the rhinoceros-like herbivore, *Coryphodon*; *Phenacodus*, the archaic ungulate that represents the ancestry of modern even-toed ungulates; and *Hyracotherium*, the earliest known horse.

By Middle Eocene time, Lake Gosiute had expanded in size until it occupied the entire Green River Basin, and the Green River Formation, lake sediments up to 1500 meters (5000 feet) thick, accumulated. The

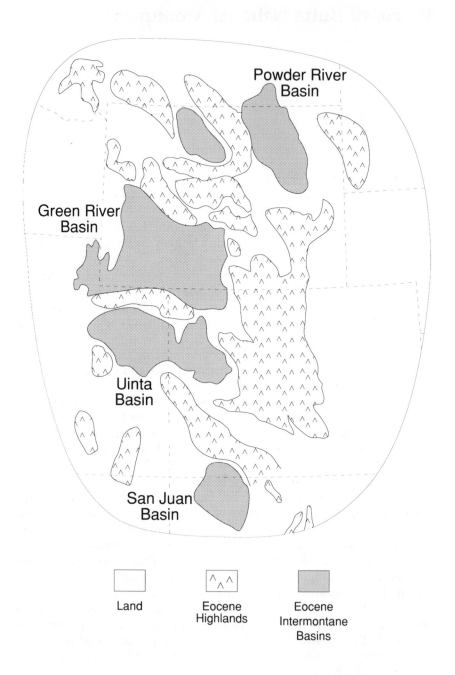

Figure 41. Intermontane basins flanked by highlands in the Central Rockies in Eocene time. The geological activity was centered in what is now Wyoming and Colorado.

expansion of the lake indicates that precipitation was considerably higher than in the Early Eocene, a significant change in climate. That this change was regional is witnessed by extensive lake sedimentation in all the intermontane basins in the Central Rockies.

The Green River Formation comprises thin, horizontally-bedded and varved shales in muted tones of tan, green, and gray. It is richly fossiliferous, renowned world-wide for beautifully preserved fossil fish. The fossils are abundant, and many different sizes and species are represented. They are found along the horizontal bedding planes throughout the Green River shale, but some horizons are much more prolific than others. Fish are the best-known fossils; frogs, birds, insects, and plant remains have also been found, each fragile organism delicately preserved by a gentle rain of muds and clays.

The Green River Formation is famous for another reason. It is one of the best known examples in the world of oil shales, fine-grained sediments containing abundant kerogen, organic material that can be distilled into petroleum. The oil shales are varved, with thin alternating light and dark bands. Each couplet, comprising a light and dark band, is thought to represent the deposits of a single year: the light-colored layers were deposited early in the season, in the spring when the water in the lake was relatively deep; over the summer as water evaporated and the lake became shallow, algal mats developed, and the dark, organic-rich layers of sediment were deposited. The layers are dark because they contain the so-called oil, the remains of the cyanobacteria that flourished on the shallow lake bottom.

Alluvial deposits began to form again in the Late Eocene; Lake Gosiute had dried and disappeared, the climate had become drier. The Bridger Formation of red sandstone beds is distinguished from the Wasatch Formation by its much higher content of volcanic ash (derived from Absaroka volcanoes [Site 13]), by interbeds of limestone that developed in ephemeral lakes, and by a different assemblage of fossil mammals. Particularly common in the Bridger Formation are the uintatheres, very large rhinoceros-like animals characterized by very long skulls bearing bizarre bony protuberances. *Uintatherium*, for example, had six variously-placed horns on the head and prominent, enlarged upper canines that extended below the level of the lower jaw and were protected by special flanges of bone from the lower jaw.

A small, subsidiary basin was present on the western margin of the Green River Basin and occupied by a lake called Fossil Lake.[18] This lake was considerably deeper than Lake Gosiute, but it was smaller and not as long-lived. The pattern of sedimentation and the sediments themselves—fluvial and alluvial sediments of the Wasatch, lacustrine shales

---

[18]The Eocene Fossil Lake of southwestern Wyoming is not to be confused with the Pleistocene Fossil Lake of central Oregon [Site 8].

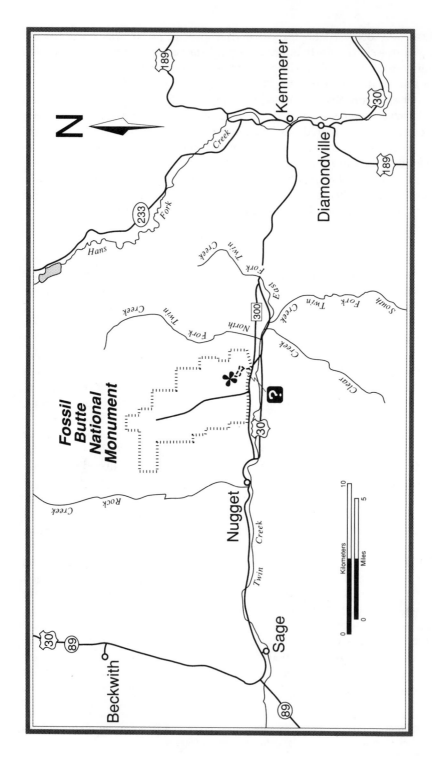

Figure 42. Location of Fossil Butte National Monument, Wyoming.

of the Green River, and alluvial sediments of the Bridger formations—are identical to those of the Green River Basin.

Fossil Lake is the focus of Fossil Butte National Monument. It highlights the unique fossils of the Green River Formation, exposed as a prominent ridge overlooking the valley. The Wasatch Formation on the valley floor and the lower slopes of the ridge is fossiliferous within the monument, although sparsely so. The ridge of Green River shale at Fossil Butte contains a particularly fossil-rich layer, a band almost 50 centimeters (20 inches) thick of carbonate-rich shale near the top. For more than 100 years, this horizon has been quarried, and the fossil fish— perch, herring, gar, etc.—sold commercially. Quarrying is no longer allowed; the best fossils are housed indoors at the visitor center, among them, *Knightia*, the state fossil of Wyoming.

DIRECTIONS: Access to Fossil Butte National Monument via United States Route 30 is 19 kilometers (12 miles) west of Kemmerer (Figure 42). The visitor center is 5 kilometers (3 miles) north of United States Route 30 on County Route 300.

PUBLIC USE: Season and hours: Fossil Butte National Monument is solely dedicated to the preservation of a unique fossil site. It is open to the public year round, although outdoor use may be restricted by snow cover. **Fees:** None. **Recreational activities:** Back-country hiking and picnicking are allowed on site. **Restrictions:** Collecting of fossils is prohibited.

EDUCATIONAL FACILITIES: Visitor Center: A new visitor center, completed in 1990, provides expanded space to house an excellent display of fossils from the Green River shale. It includes a demonstration area to show the special techniques required to prepared the fossil fish. **Visitor Center hours:** The visitor center is open year round: daily; from 8:00 A.M. to 4:30 P.M.; with extended hours May 15 to September 15 to 5:30 P.M. **Fees:** None. **Trails:** A 4 kilometer (2.5 mile) self-guiding trail leads to the site of an historical quarry, from which many fossil fish were removed (trailhead near the monument entrance at the site of the former visitor center). The trail is relatively easy to walk, only becoming steep at the ridge where the quarry was excavated. Remains of the fossils can still be seen and one can smell the oil in the dark, organic-rich layers. **Staff programs:** Park rangers give guided walks (scheduled on weekends) and campfire programs (every Monday evening) during the summer. Hands-on fossil preparation programs are available at the visitor center for children as part of the Junior Ranger Program.

FOR ADDITIONAL INFORMATION: Contact: Superintendent, Fossil Butte National Monument, P. O. Box 592, Kemmerer, Wyoming 83101, (307) 877–4455. **Read:** (1) Boyer, Bruce W. 1982. Green River laminates: does the playa-lake model really invalidate the stratified-lake model? Geology, volume 10, pp. 321–324. (2) Grande, Lance. 1984. Paleontology of the Green River Formation, with a Review of the Fish Fauna. Second Edition. The Geological Survey of Wyoming, Bulletin 63. (3) Jackson, Richard W. 1980. The Fish of Fossil Lake: The Story of Fossil Butte National Monument. Jensen, Utah: Dinosaur Nature Association in cooperation with the National Park Service.

16. Petrified Forest
Theodore Roosevelt National Park

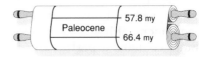

Paleocene — 57.8 my / 66.4 my

Medora, North Dakota

In North Dakota, in the Badlands of the Little Missouri River, a thick blanket of sediment has been carved and exposed to view. The sediments are part of the Fort Union Formation, deposited in earliest Tertiary time some 60 million years ago. They record a time of transition when global events coincided with regional tectonic activity—the epicontinental seas were drying as sea levels world-wide fell, and volcanism was increasing on the west coast of North America.

The Fort Union Formation is a deposit of sands and silts: soft tan, gray, and charcoal in color; interbedded with thick coal seams; punctuated by brilliant red-orange scoria horizons, clay layers that were baked by the heat of burning coal seams. The Fort Union Formation is widely distributed on the north-central plains, its lateral extent coinciding approximately with the margins of the Williston Basin that had developed in Paleozoic time (see Stonewall Quarry Park [Site 1]).

The Fort Union Formation was deposited as an ancient alluvial fan, a complex association of river, lake, and swamp deposits. Deposition began early in Paleocene time. As the Western Interior Seaway, so persistent during Late Cretaceous time (see Dinosaur Provincial Park [Site 2]), receded and dried, sediments derived from the mountains in the west were carried farther and farther inland, into the lakes and swamps that were the remnants of the sea. The lowlands became clogged with sediments, forests became well established, and coal swamps flourished. Simultaneously, along the western margin of North America, mountain building and volcanism were beginning anew. Large quantities of silica-rich ash extruded by the volcanoes were carried eastward downstream by rivers and incorporated in the sediments on the alluvial plain.

The forests on the Fort Union landscape were repeatedly inundated by flood waters. Trees died, but many remained in growth position. Sediment settled out of the water and buried the lower portions of the trees; the upper, unburied portions were exposed to air and rotted away. New growth of forests developed on the sediments that had buried the old, only to become choked and buried in turn. The groundwater was

146

rich in dissolved silica, and it petrified the trees as it percolated through the sediments. In this way, successive forest growth in Paleocene time was transformed into the fossil forests of Theodore Roosevelt National Park. Now the petrified trees are being eroded out of the enclosing matrix and are toppling over, many revealing large portions of their root structures still intact.

Coal swamps, recurrent features of alluvial plains, are notorious for their oxygen-depleted waters. The analogy with modern environments is compelling: the organic matter that accumulated in Paleocene swamps did not decay; rather, it was transformed into a soft, woody coal called lignite.

The fossil flora preserved in the Fort Union Formation in Theodore Roosevelt National Park consists of petrified trees and coal-swamp plants. Among the trees are *Metasequoia*, the dawn redwood, and deciduous hardwoods: willow, palm, ash, maple. The coals yield cycads, ferns, ginkgos, algae, fungi, and leaves of deciduous trees. Taken together, the flora indicates that the climate was subtropical to warm-temperate, evoking the southeastern United States today. The presence of extensive moist lowlands is confirmed, but many of the flowering plants in the flora suggest that the climate was seasonal and, in the upland areas away from the floodplains and swamps, decidedly dry. Certainly the interior of North America did not receive the same quantity of rainfall as coastal regions to the west did, and subtropical rainforests did not persist for as long (compare John Day Fossil Beds National Monument [Site 7]).

The fossils in Theodore Roosevelt National Park are concentrated in the Petrified Forest. There one walks as in a forest among the petrified trees, many in growth position, many now fallen over with their roots exposed; walking on a litter of petrified wood as animals once walked on a litter of leaves on the forest floor (Figure 43). The trees occur in several stratigraphic levels, the cycle of growth, burial, and preservation repeated by fresh growth of trees.

In other areas of the park, one can see coal seams and scattered petrified trees. Particularly striking are the bands of scoria, fossils in an indirect sense for they indicate only that coal was once present. The coals of the Fort Union Formation are a tremendous reserve of fossil fuel, readily recoverable, comparatively low in sulfur, even though low in quality and high in waste ash when burned.

DIRECTIONS: Access to Theodore Roosevelt National Park (South Unit) is via Interstate 94 west of Belfield (Figure 44); take Exit 6 or 7 (depending on direction of travel) to the historical town of Medora and to the main entrance to the park on the west side of the town. The Medora Visitor Center is 0.3 kilometer (0.2 mile) north of the main Medora road. South Loop Road is a 57 kilometer (36 mile) scenic route through the park offering many opportunities to view the badlands and see coal seams, scoria horizons, and isolated petrified trees. Exit 8 on Interstate 94 allows access to the Painted Canyon Visitor Center.

Figure 43. The Petrified Forest in Theodore Roosevelt National Park, North Dakota. The photograph, overlooking the badlands, illustrates in the foreground the petrified wood that has been exposed by weathering.

PUBLIC USE: Season and hours: Theodore Roosevelt National Park is open year round although access to the Petrified Forest and other exposures of fossils may be limited in winter by snowfall. **Fees:** May 1 to September 30: $3.00/vehicle or $1.00/person (entering park as pedestrian or on commercial carrier) daily or $25.00 annual Golden Eagle Permit (inquire about other park permits). **Food service:** Restaurants and stores are open in Medora only during the summer season. **Recreational activities:** A full range of park activities is available including hiking (on trail or in the back country), horseback riding, picnicking, camping (for which a fee is charged from mid-April through October 30), and back-country camping (free permit required). An automobile tour along the scenic South Loop Road, and historical aspects of the park are also featured. **Handicapped facilities:** The visitor centers, campground, and picnic area are accessible by wheelchair. **Restrictions:** Collecting of fossils is prohibited.

EDUCATIONAL FACILITIES: Visitor Center: Theodore Roosevelt National Park (South Unit) has two visitor centers. The Medora Visitor Center is the primary visitor center and includes a small display of petrified wood; the Painted Canyon Visitor Center provides additional information. Each features natural history displays, with the cultural aspects of the park highlighted at the Medora Visitor Center. **Visitor Center hours:** Medora Visitor Center is open year round: daily; from 8:00 A.M. to 4:30 P.M.; with extended hours early June to early September to 8:00 P.M.; except closed New Year's Day (January 1), Thanksgiving Day, and Christmas Day (December 25). Painted Canyon Visitor Center is open mid-April to late October: daily; from 8:30 A.M. to 4:30 P.M.; with extended hours early June to end of August from 8:00 A.M. to 8:00 P.M.; except closed Thanksgiving Day. **Fees:** None. **Bookstore:** Medora Visitor Center has a small selection of books, maps, and natural history items of interest in the local area. **Trails:** One trail, the Petrified Forest Loop Trail, leads to

148

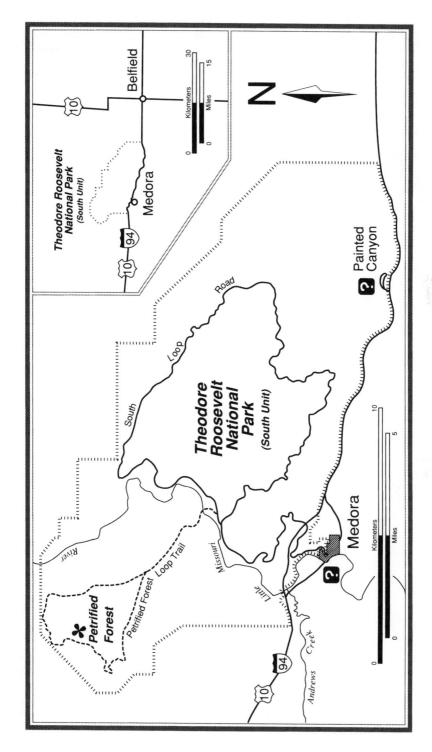

Figure 44. Location of the Petrified Forest in Theodore Roosevelt National Park (South Unit), North Dakota.

the Petrified Forest. It is a hiking and horseback riding trail that begins at Peaceful Valley Ranch on South Loop Road (north of Medora) and leads through the Petrified Forest. Although the distance is long (approximate round trip distance of 26 kilometers or 16 miles, but distance varies depending upon trail selection), the trail is easy to walk. It is marked, but alertness is required because the marker posts are repeatedly knocked over by the resident bison herd. Interpretive information and trail maps should be obtained in advance at the Medora Visitor Center. **Staff programs:** Park interpretive programs are available only in the summer. Included are daily ranger-guided walks, cabin talks, and campfire talks. Special hikes can be scheduled to the Petrified Forest.

FOR ADDITIONAL INFORMATION: Contact: Superintendent, Theodore Roosevelt National Park, Medora, North Dakota 58645, (701) 623–4466.

17. Badlands National Park

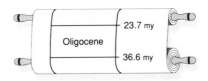

Oligocene
23.7 my
36.6 my

Interior, South Dakota

Exposed in the White River Badlands in Badlands National Park are sediments of Late Eocene and Oligocene age deposited by ancient rivers that flowed eastward from the Rocky Mountains and the Black Hills. Collectively called the White River Group, but subdivided into three formations: Chadron, Brule, and Sharps; 37 to 23 million years old, they are the thickest and most widely distributed sediments in the area. The wealth of unique data derived from them documents the complex history of Late Eocene and Oligocene time of the Great Plains. Adjacent to the park are even younger sediments, Early Miocene in age.

Two distinct kinds of fossils occur in the sediments of Badlands National Park: bones and paleosols. The bones are remains of extinct animals that lived in and along the rivers, and in adjacent upland areas; the paleosols are fossil soils, 87 identified within the park, the highest known concentration of fossil soils in the world.

The formation of paleosols is dependent upon the chemistry of iron, which has a strong affinity for organic matter. When iron is present in sediment and groundwater, it tends to become oxidized by and concentrated within the organic matter. Like modern soils, the soils that developed on Oligocene land surfaces had a high proportion of organic matter. Over time the iron oxides became concentrated within the organic-rich strata and now impart a distinctive red color, sometimes brown or tan tinged with red, to the ancient soils; the paleosols impart a distinctive banding to the sediments.

Among the 87 paleosols, distinct, repeated groups of soils are recognized, each group corresponding to the particular environment within which the soil formed. On the basis of the soils, the vegetation can be reconstructed and the climate inferred. In this way, it is possible to document the climatic changes that took place over some 15 million years on the Great Plains.

The evidence from paleosols indicates that, late in the Eocene, a dense, closed-canopy forest blanketed what is now the Central Great Plains. The climate was subtropical, hot and humid with mean annual temperatures perhaps as high as 25 to 30 degrees Celsius (about 80 degrees Fahrenheit). The paleosols change upward in the sedimentary sequence, an indication that, over time during the Late Eocene and into the Oligocene, the vegetation changed. The soils indicate that dense forest gave way to woodlands and open meadows, and eventually to savannahs. Grasses and herbs grew in the open areas; forests were restricted to lowlands and river valleys. It is clear that the climate, moist and subtropical late in the Eocene, became cooler and increasingly drier in the Oligocene. Mean annual temperatures during Oligocene time were probably about 15 degrees Celsius (60 degrees Fahrenheit), perhaps fluctuating but not declining markedly over that time. The amount of rainfall, however, declined dramatically over time, and the annual distribution became more markedly seasonal. Soils at the top of the Brule Formation indicate that grasslands ultimately replaced the savannahs. Thus late in the Oligocene, seasonal droughts were probably common, the climate semi-arid and cool-temperate.

Oligocene sediments in Badlands National Park are rich in vertebrate fossils, primarily mammals. The vertebrate fossils corroborate the interpretations based on paleosols. The earliest mammals known in the White River Group are primitive horses, tapirs, rhinoceroses, and titanotheres; the odd-toed ungulates, browsers eating the soft, lush vegetation of the subtropical forests. During Oligocene time, grazing and cursorial mammals, such as the sheep-like oreodont, *Merycoidodon*, and the three-toed horse, *Mesohippus*, evolved to occupy the grassy, open savannahs; the small deer-like artiodactyls, such as *Leptomeryx*, were among the most common inhabitants of the savannahs. For the first time in the history of mammals, even-toed ungulates, the artiodactyls, became more common than the odd-toed ungulates. The transition seen among the fossil mammals in Badlands National Park foreshadows the modern condition, recording the modernization of mammals in response to changing climates and vegetation.

The sediments of the Chadron and Brule formations—almost 200 meters (650 feet) thick, variegated pastel white, gray, pink, and red sands, silts, and clays, and rich in volcanic ash—are exposed throughout Badlands National Park. They are made conspicuous by their red horizontal striping. Some beds, especially in the upper part of the sequence, are composed entirely of volcanic ash; the Rockyford Ash, for example, is 10 meters (35 feet) thick in places. The sediments are unconsolidated and poorly cemented throughout, but the fossils they contain are very well preserved thanks to the silica derived from volcanic ash.

Vertebrate fossils are common, occurring as entire skeletons or as isolated bones that are well disseminated throughout the sediments. On the one hand, fossils are abundant; on the other, many layers are devoid of fossils. There are some bonebeds; the titanothere beds in the Chadron Formation are areas of unusually high concentrations of bone. Seeing

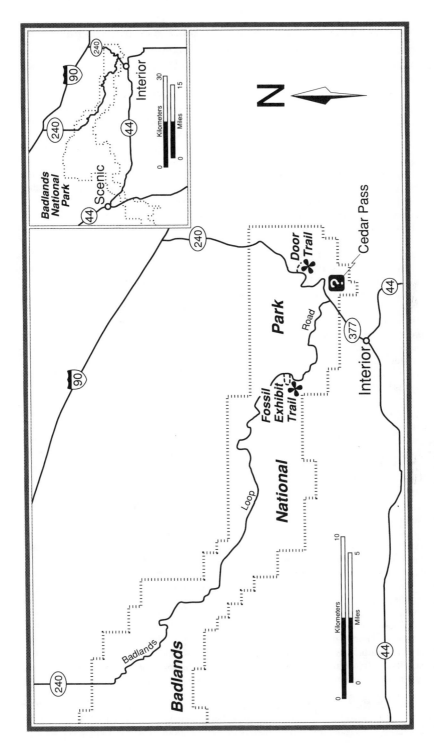

Figure 45. Location of Badlands National Park, South Dakota.

153

fossils in the badlands, however, is very much a chance phenomenon dependent upon erosion to expose the fossils. Nonetheless, erosion continues to expose fossils everywhere in the badlands.

DIRECTIONS: The main access to Badlands National Park is to the northern and eastern portions, the most developed parts of the park (Figure 45). Access is via Interstate 90; eastbound take Exit 110 (at Wall) and follow Badlands Loop Road (State Route 240) to the Pinnacles Entrance Station. From there the Badlands Loop Road winds through the park to the Cedar Pass Visitor Center. Continue on the Badlands Loop Road north to Interstate 90 at Exit 131. The total distance from Exit 110 to Exit 131 is approximately 65 kilometers (40 miles). Westbound on Interstate 90 take Exit 131 and follow Badlands Loop Road in reverse. Access from State Route 44 is via Scenic (1.6 kilometers or 1 mile east of Scenic turn northeast onto the gravel road to Sage Creek Rim Road and Badlands Loop Road) or via Interior to State Route 377 which leads to the Cedar Pass Visitor Center. Access to the south unit of the park is by various routes leading from State Route 44.

PUBLIC USE: Season and hours: Badlands National Park is open year round and access to trails featuring fossils is possible at any time. **Fees:** May 1 to September 30: $3.00/vehicle or $1.00/person (age 17 to 61 inclusive) daily, or $25.00 annual Golden Eagle Permit (inquire about other park permits). **Food service:** There is a restaurant at the concessioner-operated lodge located adjacent to the Cedar Pass Visitor Center. **Recreational activities:** Hiking trails, some of which are well-signed, self-guiding trails; camping (for which a fee is charged in summer; there is no fee charged for use of the primitive campground), and picnicking are available. **Handicapped facilities:** Visitor centers, some trails (the Fossil Exhibit Trail, the Window Trail, and the approach to the first lookout on the Door Trail), and some facilities at Cedar Pass Campground are accessible by wheelchair. **Restrictions:** Collecting of fossils is prohibited.

EDUCATIONAL FACILITIES: Visitor Center: Cedar Pass Visitor Center highlights the fossils of the park and features natural history exhibits, demonstrations, audiovisual program, and fossil displays. The White River Visitor Center in the South Unit concentrates on the human history of the badlands. **Visitor Center hours:** Cedar Pass Visitor Center is open year round: daily; from 8:00 A.M. to 4:30 P.M.; with extended hours in the spring (April and May) and fall (September and October) to 5:00 P.M., and in summer (June, July, and August) from 7:00 A.M. to 8:00 P.M.; except closed New Year's Day (January 1), Thanksgiving Day, and Christmas Day (December 25). White River Visitor Center is open early June to late August: daily; from 9:00 A.M. to 5:00 P.M. **Fees:** None. **Bookstore:** Cedar Pass Visitor Center offers a selection of books, maps, and gift items on natural and human history of the badlands region. **Trails:** Two trails in the park highlight fossils. The Fossil Exhibit Trail is a short, easy-to-walk 0.4 kilometer (0.25 mile) loop that has seven displays of badlands fossils (replicas) under protective plastic bubbles. Along the self-guiding Door Trail (1.2 kilometers or 0.75 mile return), the best introduction to the badlands and the geological history they represent, one can see several of the many paleosols that are known in the badlands. The Rockyford Ash is prominent along the trail. The Fossil Exhibit Trail is located 8.1 kilometers (5 miles) west of the Cedar Pass Visitor Center on Badlands Loop Road; the Door Trail 3.8 kilometers (2.4 miles) northeast. Remember that when hiking almost anywhere in the badlands, it is possible to see freshly eroded and exposed fossils. **Staff programs:** Interpretive programs which include guided hikes, talks, and fossil preparation demonstrations are available only from early June through Labor Day.

FOR ADDITIONAL INFORMATION: Contact: Superintendent, Badlands National Park, Interior, South Dakota 57750, (605) 433–5361. **Read:** Retallack, Greg J. 1983. Late Eocene and Oligocene Paleosols from Badlands National Park, South Dakota. Geological Society of America, Special Paper 193.

18. The Mammoth Site of Hot Springs, South Dakota, Inc.

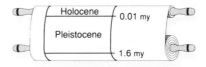

Hot Springs, South Dakota

The last great ice sheet of the Pleistocene ice ages, the Wisconsinan ice sheet, at its maximum covered much of North America. It extended southward to about 48 degrees North latitude in the interior of the continent (Figure 31). Along its southern margins where the climate was cool and moist, a unique arctic grassland developed and served as home to a wide variety of animals. Grasses, shrubs, and bushes predominated; trees were scarce. Among the animals was the Columbian mammoth, *Mammuthus columbi.* Today, in the Black Hills, a unique deposit yields a high concentration of mammoth bones, bones of 100 animals or more.

The deposit at Hot Springs consists of silts and clays that accumulated within and eventually filled a sink-hole. In Pleistocene time, limestone rocks that are present below the surface of the earth in the Black Hills area were being dissolved by groundwater to form caverns. The rocks that were the roof of one such cave collapsed, and a steep-sided hole on the surface formed, a sink-hole which filled with water and began to fill with sediments. Meanwhile, it served as a watering hole, attracting thirsty animals. Among them were mammoths that came to drink or swim. Those that fell or ventured into the water became trapped, unable to climb out because the walls of the sink-hole were steep and slippery.

The Mammoth Site is unlike other death-trap accumulations, for it did not concentrate bones of carnivores (as at Cleveland-Lloyd Dinosaur Quarry [Site 39] and Hancock Park [Site 43]). It appears, therefore, that the carcasses of drowned mammoths sank in relatively deep water; had they become mired in shallow water and mud, scavenging carnivores would surely have been attracted to the site.

In 1974, a low hill on the outskirts of Hot Springs impeded development of a housing project, and in the process of leveling the hill, some mammoth bones were discovered. Today a visitor center covers the area, protecting the fossils, and affording visitors and researchers alike unique insight into the biology of an extinct animal.

Inside the visitor center, the yellow silts and clays that infilled the

Figure 46. Excavation at the Mammoth Site of Hot Springs, South Dakota, Inc. Inside the visitor center one sees mammoth bones preserved *in situ*. (Photograph by Brian Cluer, Mammoth Site of Hot Springs, South Dakota, Inc.)

sink-hole are immediately striking. Within the sediment, lying where they were found, unearthed to provide good visibility, are bones: skulls, tusks, teeth, shoulder blades, leg bones, entire skeletons. Walkways lead over and around the fossiliferous deposit and allow visitors to see how the bones were preserved and how they are being excavated (Figure 46).

Excavation at the Mammoth Site is an on-going endeavor. The sediment that is removed from around the mammoth bones is washed through a screen in search of the bones and shells of small animals that coexisted with the mammoths. More than 25 species of vertebrates are known from the sink-hole. Most are rodents, such as ground squirrels, prairie dogs and gophers, but unusual and rare species are present: extinct giant short-faced bear (*Arctodus simus*) and peccary. None are as abundant or spectacular as the Columbian mammoth.

The Mammoth Site of Hot Springs, South Dakota, Inc., a non-profit organization, was formed by a citizens group in Hot Springs in 1975 in response to the discovery of the unique fossil site. Its objectives insure that visitors can participate in a unique endeavour: protection, research, and contribution to scientific knowledge and public education. The importance of the site has been recognized by its designation as a United States National Natural Landmark.

DIRECTIONS: The Mammoth Site is located within the town of Hot Springs. Access via United States Route 18 Truck Bypass is 1.8 kilometers (1.1 miles) west of the east exit, or 1.2 kilometers (0.75 mile) southeast of the west exit from United States Route 18 (Figure 47).

PUBLIC USE: Season and hours: The Mammoth Site of Hot Springs, South Dakota, Inc. is entirely enclosed within a large visitor center designed to provide visitors with good access to view fossils. Access is by way of tours, which are available as frequently as demand requires and staffing permits (approximately every 10 minutes in summer), after which time visitors can spend as much time as they wish on site. It is open year round: daily; April 1 to May 15, September 1 to October 31, Monday to Friday from 9:00 A.M. to 5:00 P.M., Saturday and Sunday from 11:00 A.M. to 5:00 P.M.; May 15 to August 31 from 8:00 A.M. to 8:00 P.M.; November 1 to March 31 tours at 11:00 A.M., 1:00 P.M., 3:00 P.M. or by appointment. **Fees:** Adults $3.95 (over 60 years $3.50), 6-12 years $1.95, under 6 years free. **Food service:** Restaurants and stores are available in Hot Springs. **Handicapped facilities:** The Mammoth Site is easily accessible by wheelchair. **Recreational activities:** None on site. A full range of recreational activities is available in Hot Springs and surrounding area. **Restrictions:** Collecting of fossils is prohibited.

EDUCATIONAL FACILITIES: Visitor Center: The visitor center has displays and demonstrations in addition to the excavation itself. The local geology and the formation of the deposit is documented; the demonstrations include washing of sediment to search for small fossils. **Visitor Center hours:** The visitor center is open during the Mammoth Site hours of operation. **Fees:** The above listed fees are all inclusive. **Bookstore:** A small bookstore and gift shop sells souvenir items, books that feature mammoths and the ice ages, and items of local interest. **Tour guide:** All visitors are introduced to the Mammoth Site by a guide who takes visitors through the site explaining details of the geology of the deposit, the biology of the mammoth, excavation and preparation of fossils, and answering questions. **Special group activities:** The Mammoth Site is one of the stops on several of the field trips run by the Trailside Museum, Fort Robinson State Park (Crawford, Nebraska 69339, (308) 665–2730). An outreach program to schools in the area is now in place, providing tours and hands-on activities. Special group activities can also be arranged by contacting the Mammoth Site directly. **Note:** The Earthwatch Program offers participation in the excavation of the Mammoth Site.

FOR ADDITIONAL INFORMATION: Contact: The Mammoth Site of Hot Springs, South Dakota, Inc., P. O. Box 606, Hot Springs, South Dakota 57747, (605) 745–6017. **Read:**(1) Agenbroad, Larry D., Jim I. Mead, and Lisa W. Nelson (editors). 1990. Megafauna and Man: Discovery of America's Heartland. Hot Springs, South Dakota: The Mammoth Site of Hot Springs, South Dakota, Inc. Scientific Papers, Volume 1. (2) Laury, Robert L. 1980. Paleoenvironment of a late Quaternary mammoth-bearing sinkhole deposit, Hot Springs, South Dakota. Geological Society of America Bulletin, Part I, volume 91, pp. 465–475. (3)

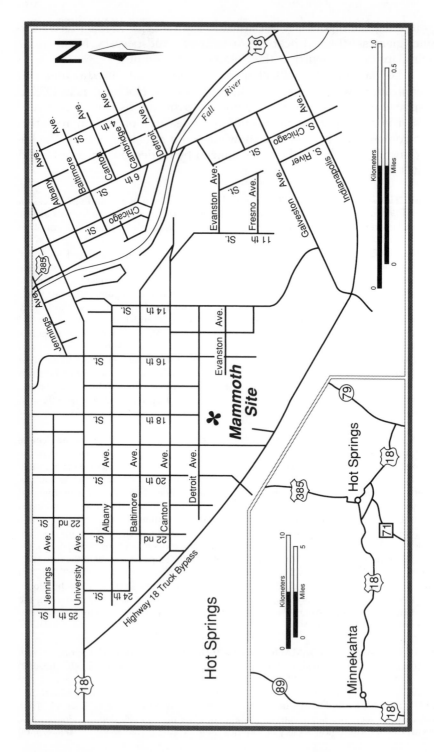

Figure 47. Location of the Mammoth Site of Hot Springs, South Dakota, Inc., Hot Springs, South Dakota.

Martin, Paul S., and Richard G. Klein (editors). 1984. Quaternary Extinctions: A Prehistoric Revolution. Tucson, Arizona: The University of Arizona Press. (4) Sutcliffe, Antony J. 1985. On the Track of Ice Age Mammals. Cambridge, Massachusetts: Harvard University Press.

19. Agate Fossil Beds National Monument

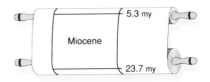

Harrison, Nebraska

Miocene sediments on the Great Plains of North America are exposed high on hilltops and along the crests of ridges. They are all that remain of what was once an almost continuous cover of sediments in the western interior—sediments carried by streams from the mountains in the west and filling in the lowlands.

The Great Plains had been the site of deposition for the first 45 or 50 million years of Tertiary time. The waters of the Western Interior Seaway had receded at the end of Cretaceous time, and sediment-laden streams began to flow across the lowlands. Layers of sediments, like blankets, were laid down; but by Miocene time, the western interior had been filled in. Then, over the last 15 million years, erosion began to reshape the Great Plains, carving the present landscape. Rivers that once deposited sediments now cut their way down through the previously deposited strata and carried the material away to be deposited elsewhere. Slices of the sediments that were laid down so long ago remain along river valleys and on hillsides.

The climate during the Miocene in the western interior of North America was dry and temperate, and some areas experienced arid conditions. In addition, seasonality in temperature and rainfall was pronounced. Grasses thrived and diversified in such conditions, and open grasslands replaced the woodlands and forests of earlier times (see Badlands National Park [Site 17]).

The Miocene sediments, deposited on the Great Plains some 20 million years ago, are represented in western Nebraska by the Harrison Formation: a layer of fluvial sands and silts, sediments deposited by sluggish, intermittent streams; and an overlying layer of eolian or wind-blown sands. The former contain the bones of mammals; the latter, the infilled burrows of an extinct terrestrial beaver (Figure 48).

The mammals that are represented by bones found at Agate Fossil Beds National Monument were new. They had evolved strategies that enabled them to survive on a diet of grasses and on open grasslands that had few trees for shelter or protection. Herbivores developed both elon-

Figure 48. *Daemonelix*, the corkscrew shaped burrow of an extinct terrestrial beaver. The burrow is weathering out of a hillside in Agate Fossil Beds National Monument. The book in the foreground for scale is approximately 20 centimeters (8 inches) long.

gate, ever-growing teeth and specialized digestive systems; the former to chew abrasive grasses, the latter to process large quantities of the nutrient-poor fodder. Many of the Miocene mammals were agile and swift runners. *Menoceras*, for example, was a miniature grazing rhinoceros; *Stenomylus*, a small, swift member of the camel family.

The eolian sands of the Harrison Formation preserve a most unusual kind of trace fossil: *Daemonelix*, giant corkscrew burrows that were infilled with sediment and preserved. The burrows were an enigma to paleontologists until skeletons of the large non-aquatic beaver, *Palaeocastor*, were found preserved at the bases of some of the corkscrews.

The trace fossils are an important source of information about the behavior of *Palaeocastor*. Analysis of *Daemonelix* has produced interesting results: apparently 50 percent of corkscrews turn clockwise, 50 percent counterclockwise. The implication is that the direction of digging was a random choice by each animal. There is no way, however, to determine if any one beaver always dug in the same direction. Perhaps the direction a beaver dug was random every time the animal made a choice; perhaps every animal dug in only one direction, a right-handed or left-handed beaver, and 50 percent of the beavers were right-handed and 50 percent were left-handed. It is also clear from mapping the distribution of burrows that they were arranged in towns as are the burrows of living prairie

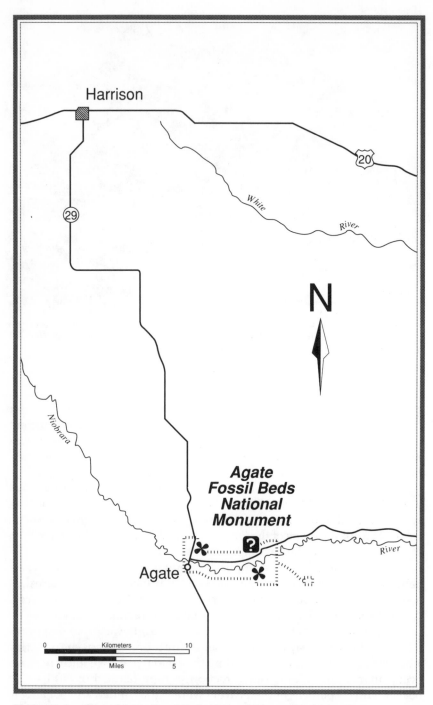

Figure 49. Location of Agate Fossil Beds National Monument, Nebraska.

dogs; the analogy suggests that *Palaeocastor* had a similar, well-developed social structure.

Agate Fossil Beds National Monument is situated in the valley of the Niobrara River. Two prominent erosional remnants, Carnegie Hill and University Hill, dominate the southern margin of the valley. The bones of mammals occur in small bonebeds near the crests of the hills. The corkscrew fossils can be seen high on the ridge along the northern margin of the valley.

DIRECTIONS: Access to Agate Fossil Beds National Monument is via State Route 29 (Figure 49): from the north from Harrison and United States Route 20 the distance is 35 kilometers (22 miles); from the south from United States Route 26 near Mitchell the distance is 55 kilometers (34 miles). An unpaved park road provides access from State Route 29 to the park visitor center (5 kilometers or 3 miles) and continues eastward to Marsland.

PUBLIC USE: Season and hours: Agate Fossil Beds National Monument is devoted to the preservation of a unique fossil area. It also preserves native prairie and houses the Captain James H. Cook Collection of Indian artifacts. It is open year round, and the fossils can be seen at any time. Development plans for the monument promise additional displays and observation of *in situ* fossil preparation. **Fees:** None. **Recreational activities:** Picnic facilities are available. **Restrictions:** Collecting of fossils is prohibited.

EDUCATIONAL FACILITIES: Visitor Center: The visitor center features natural history displays that highlight the fossils of the monument. **Visitor Center hours:** The visitor center is open year round: weekends; from 8:00 A.M. to 5:00 P.M.; with daily opening Memorial Day to Labor Day from 8:30 A.M. to 5:30 P.M.; except closed New Year's Day (January 1) and Christmas Day (December 25). **Fees:** None. **Trails:** There are two developed trails each leading to fossils exhibited *in situ*. One trail (1.6 kilometers or 1 mile, one way) leads from the visitor center to excavation sites on Carnegie Hill and University Hill. Fossils can be seen protected beneath plastic bubbles. Another trail (0.8 kilometer or 0.5 mile, one way), this one near the park entrance, leads to the *Daemonelix* sites. The trails are short, and the walking easy. **Tour guide:** Guided tours are available for special groups and school tours by prior arrangement. **Special group activities:** The Trailside Museum, Fort Robinson State Park (Crawford, Nebraska 69339, (308) 665–2730) has incorporated stops at Agate Fossil Beds into its public field trips.

FOR ADDITIONAL INFORMATION: Contact: Superintendent, Agate Fossil Beds National Monument/Scotts Bluff National Monument, P. O. Box 27, Gering, Nebraska 69341, (308) 436–4340. **Read:** Martin, Larry D., and Debra K. Bennett. 1977. The burrows of the Miocene beaver *Palaeocastor*, western Nebraska, U.S.A. Palaeogeography, Palaeoclimatology, Palaeoecology, volume 22, pp. 173–193.

20. Kremmling Cretaceous Ammonite Locality

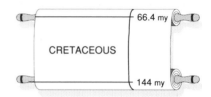

CRETACEOUS

66.4 my

144 my

Kremmling, Colorado

During Late Cretaceous time, the western interior of North America was flooded from the Gulf of Mexico to the Beaufort Sea by an epicontinental sea, the Western Interior Seaway. The lateral extent of the seaway fluctuated as sea levels world-wide rose and fell; in latest Cretaceous time, sea levels once again rose and the Western Interior Seaway expanded. Mountains to the west, in the Cordillera, were still rising, producing a continuous supply of sediment that was carried eastward by rivers and deposited in the seaway. The sediments, now the Pierre Shale (Bearpaw Shale in Canada), were fine-grained muds and clays intercalated with lenses of sands deposited near shore during storms.

The shales deposited in the Western Interior Seaway, 75 to 70 million years ago, are widely distributed and exposed across the western interior of North America. They are laterally continuous to the west with dinosaur-bearing terrestrial sediments, such as those of the Judith River Formation at Dinosaur Provincial Park [Site 2], and correlative with the Two Medicine Formation in the Willow Creek Anticline [Site 12].

At the Kremmling Cretaceous Ammonite Locality, a resistant bed of sandstone forms a prominent ridge upon which lie up to 15 meters (50 feet) of sandy shales and lenses of sandstone. More than 100 fossil species are known from these strata alone, the diverse marine invertebrate fossils representing the richness of life in the Western Interior Seaway. The coiled shells of ancient squid-like creatures, giant ammonites called *Placenticeras meeki*, are particularly striking.

Among ammonites, *Placenticeras meeki* was unusual only for its large size, for the ammonites themselves—coiled, free-swimming mollusks —were the predominant invertebrate predators in the seas of Late Cretaceous time. They were capable of active and aggressive swimming: using jet propulsion as the mechanism, they squirted water out of a body tube called a siphuncle to generate thrust. The shell provided buoyancy by being divided into many gas-filled chambers.

The mechanism of jet-propelled swimming in extinct ammonites is inferred from observations of swimming in the living *Nautilus*. The

Nautilus is equipped with a siphuncle, located beneath the head, through which it expels water at great speed and propels itself in the direction opposite to water flow. This is the normal swimming pattern in *Nautilus*; it is very efficient and can generate speeds of 70 kilometers per hour (40 miles per hour), but it has a severe limitation. The animal swims backwards! In *Nautilus*, to compensate, the siphuncle is flexible, and the animal can adjust it to eject water backward and propel itself forward. Speed is sacrificed, however, because this mechanism is less efficient. There is no doubt that *Placenticeras meeki* was able to swim in the normal backward mode, but whether it possessed the mechanism for the secondary, forward swimming cannot be determined from the fossils.

The high concentration of the ammonites in the sandy lenses of the Pierre Shale is anomalous since the animals were free-swimming and wide-ranging in the sea, but it must be explained in terms of the storms that deposited the sands. During storms, large numbers of *Placenticeras meeki* may well have been carried from the open seaway into the shallow water near shore, and many individuals may have been stranded high on the beaches by the storms waves. Others may well have propelled themselves into beach areas by their backward swimming as they attempted to dislodge themselves from shallow water. When the storm passed, the stranded and beached ammonites littered the shoreline.

Analogy with living *Nautilus* suggests an alternative explanation for the unusually high concentration of ammonite shells in the sandy lenses of the Pierre Shale. When an individual *Nautilus* dies, its gas-filled shell floats on the surface of the ocean. Waves and currents distribute the shells over a wide area and deposit them on beaches. Similarly, shells of dead *Placenticeras meeki* floating on the seaway may have been concentrated by storm waves. There is no clear evidence from the fossils, however, which would indicate whether the storms stranded living animals or empty shells.

Placenticeras meeki, like many species of ammonites, was sexually dimorphic. At the Kremmling Cretaceous Ammonite Locality both the large female and small male conchs are represented as fossils. They commonly occur within concretions which, when split open, reveal large, bird-bath-like molds (Figure 50). Discarded molds litter the site, catching rainfall in true bird-bath fashion and giving the area its local name, the Bird-bath Locality. Other fossils include the characteristic Late Cretaceous ammonite, *Baculites compressus*, and the bivalve mollusk, the clam-like *Inoceramus sagensis*.

DIRECTIONS: Kremmling located on United States Route 40 is 75 kilometers (46 miles) northwest of Interstate 70 (Figure 51). Directions to the Kremmling Cretaceous Ammonite Locality are readily available at the Kremmling Resource Area Office, Bureau of Land Management, P.O. Box 68, Kremmling, Colorado 80459, (303) 724–3437. Access to the site is by way of unpaved roads, and it is necessary to walk a short distance (about 1 kilometer, 0.6 mile) to the actual site from the end of the roadway.

Figure 50. Fossils at Kremmling Cretaceous Ammonite Locality. The upper photograph illustrates a broken ammonite shell viewed in cross-section; the lower, a concretion that was split to remove the ammonite shell leaving only the bird-bath-like mold as evidence of the shell.

166

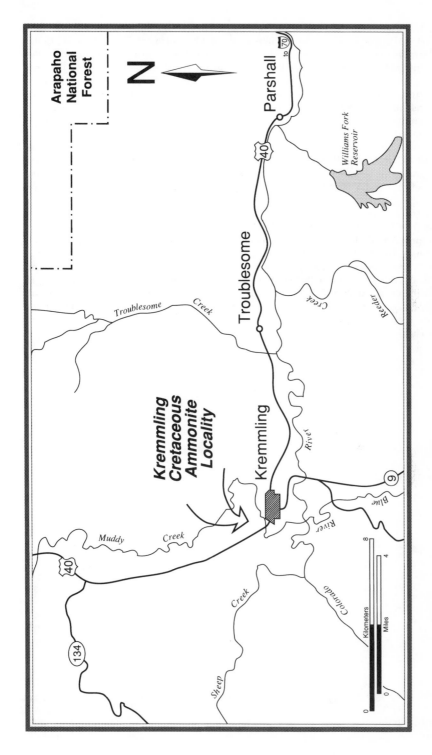

Figure 51. Location of Kremmling, Colorado, for access to Kremmling Cretaceous Ammonite Locality.

PUBLIC USE: Season and hours: The Kremmling Cretaceous Ammonite Locality is located on public land administered by the Bureau of Land Management. The site is open to the public year round, but the Bureau of Land Management is concerned about protecting the integrity of the site and requests that visitors obtain maps and directions from the Area Office before proceeding to the site. Visitors are welcome at the Area Office, and they are encouraged to visit the fossil site. Note that weather conditions, even a slight rainfall, may severely limit the access. **Fees:** None. **Food service:** Restaurants and stores are available in Kremmling. **Recreational activities:** None on site. **Restrictions:** Collecting of fossils is prohibited.

EDUCATIONAL FACILITIES: Interpretive sign: The Bureau of Land Management has designated the Kremmling Cretaceous Ammonite Locality an Area of Critical Environmental Concern and Research Natural Area. It has fenced the site and erected an interpretive sign.

FOR ADDITIONAL INFORMATION: Contact: Bureau of Land Management, Kremmling Resource Area Office, P. O. Box 68, Kremmling, Colorado 80459, (303) 724–3437. **Read:** (1) Lehman, Ulrich. 1981. The Ammonites: Their Life and Their World. English Edition. Cambridge, England: Cambridge University Press. (2) Kennedy, W. J., and W. A. Cobban. 1976. Aspects of ammonite biology, biogeography, and biostratigraphy. Special Papers in Palaeontology, Number 17. London, England: The Palaeontological Association. (3) Ward, Peter Douglas. 1988. In Search of Nautilus: Three Centuries of Scientific Adventures in the Deep Pacific to Capture a Prehistoric-Living-Fossil. New York, New York: Simon and Schuster.

21. Dinosaur Ridge

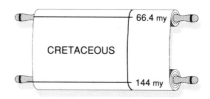

Morrison, Colorado

Near the town of Morrison, Colorado, the Dakota Hogback rises out of the earth like a fortress wall on the western outskirts of Denver. The barrier to Denver has been breached by Interstate 70, and the internal stratigraphy of the hogback is exposed. There, at the intersection of Interstate 70 and State Route 26 (Alameda Parkway) interpretive panels at a roadside geological park translate the geological story written in the layered rocks of the hogback; south on the Alameda Parkway another park, Dinosaur Ridge, focuses on dinosaur footprints.

The geological formation for which the Dakota Hogback is named is the Dakota Sandstone, a resistant cap of rock that protects the softer sedimentary rocks beneath it from erosion. During Early Cretaceous time, approximately 110 million years ago, the western interior of North America was being increasingly flooded by epicontinental seas advancing from the Beaufort Sea in the north and from the Gulf Coast in the south (Figure 52). Eventually, later in the Cretaceous, the seas would join to form the Western Interior Seaway. As the northern sea advanced southward across Wyoming, Colorado, and New Mexico and expanded eastward across the Dakotas, Nebraska, and Kansas, it deposited a layer of sands now called the Dakota Sandstone.

The Dakota Sandstone is poorly fossiliferous in the usual sense, for it yields few dinosaur bones. Nonetheless, it preserves many dinosaur footprints, including some with well preserved skin impressions; a variety of plant impressions, and many invertebrate tracks. All are evidence of life on the sandy tidal flat that separated a heavily forested coastal plain from the inland sea.

There are at least two types of dinosaurs, both bipedal, represented by the tracks at Dinosaur Ridge; one left by a small carnivore (coelurosaur), another by ornithopod herbivores. The tracks of the latter are particularly interesting because they preserve footprints of the front foot (manus) as well as the hind foot (pes), unusual and unexpected footprints for tracks of a bipedal animal. The tracks of the hind foot resemble footprints of iguanodontids (a family of large, heavy, elephantine orni-

169

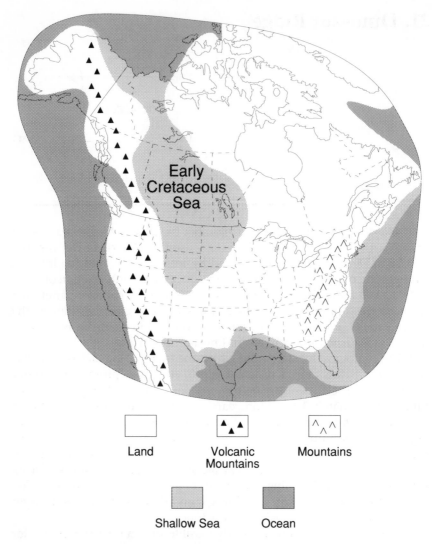

Figure 52. Western Interior Seaway forming during Early Cretaceous time.

thopods, for example, *Iguanodon*), but all other known iguanodontid tracks are strictly bipedal. The alternative interpretation, that the tracks may be those of a duck-billed dinosaur, has important implications, because hadrosaur bones are known only from rocks of Late Cretaceous age.

Trackways present a dynamic picture of dinosaurs. They tell of dinosaurs that left no bones behind: gregarious animals living in small family-like groups or in herds, migratory ones moving up and down along the shoreline of the sea, and agile carnivores stalking their prey. They give evidence of the numbers of animals present, an approximate census of the populations and the demographics of the populations. Dinoturbation, a term used to describe the trampling of sediments by dinosaurs, is a measure of the numbers of dinosaurs and their occupation of or presence in an area.

Dinosaur trackways are common in the sediments that were laid down as the Early Cretaceous epicontinental seas flooded the western interior of North America: in the Dakota Sandstone deposited on the margins of the northern sea (also exposed at Clayton Lake State Park [Site 25]), and in the coeval carbonate rocks, such as the Glen Rose Formation exposed at Dinosaur Valley State Park, [Site 28], on the margins of the Gulf Coast sea. The two environments, however, supported distinct communities of dinosaurs. The tracks in the Dakota Sandstone derive predominantly from ornithopods; theropods are represented by only one-third of the tracks. In contrast, along the Gulf Coast sea, theropod tracks are most numerous, one-quarter of the tracks attributed to sauropods, and ornithopods are rare. Ecological zonation of dinosaur communities was clearly well developed.

The area around Morrison has historical significance for vertebrate paleontologists. The Morrison Formation of Late Jurassic age is well exposed stratigraphically below the Dakota Sandstone. In 1877, the first dinosaurs from the Morrison Formation were discovered here, and the dinosaur rush of the 1880s produced many dinosaurs which were shipped to museums in the east. The old quarries are still visible, and bones are still being found.

DIRECTIONS: At the point where Interstate 70 breaches the Dakota Hogback is a well-marked exit providing access to State Route 26 (Alameda Parkway). South along the Alameda Parkway, Dinosaur Ridge encompasses the hogback (Figure 53).

PUBLIC USE: Season and hours: Dinosaur Ridge is owned by the Colorado State Highway Department, but negotiations are underway to transfer ownership to the town of Morrison. Currently, Alameda Parkway breaches the hogback, and fossils are accessible right at the edge of the roadway on the east side of the hogback. The park is accessible year round but snow cover may limit what one can see. **Fees:** None. **Recreational activities:** None are currently available on site. **Food service:** Restaurants and stores are available in surrounding communities. **Restrictions:** Collecting of fossils is prohibited.

EDUCATIONAL FACILITIES: Dinosaur Ridge is a newly structured park, and the educational facilities are in various stages of planning and development. An interpretive trail is partly completed. Efforts are being made to close the road to vehicles and turn it into a wheelchair-accessible trail through the park. Long term plans include the development of a visitor facility; in the meantime, the Morrison Natural History Museum cooperates with The Friends of Dinosaur Ridge to fulfill that role. Tours and guides are available by prior arrangement with The Friends of Dinosaur Ridge. Pamphlets are available in a box outside the fence that surrounds the tracksite and at the museum. At the roadside geological park at the Alameda Parkway exit from Interstate 70, the geological history preserved in the stratigraphy of the hogback is interpreted.

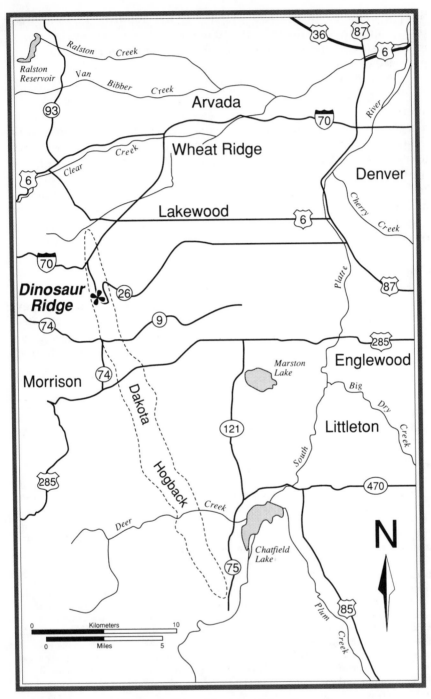

Figure 53. Location of Dinosaur Ridge, Morrison, Colorado.

FOR ADDITIONAL INFORMATION: Contact: The Friends of Dinosaur Ridge, c/o Morrison Natural History Museum, P. O. Box 564, Morrison, Colorado 80465, (303) 697–1873. **Read:** (1) Colbert, Edwin H. 1984. The Great Dinosaur Hunters and Their Discoveries. New York: Dover Publications Inc. (2) Currie, Philip J., Gregory C. Nadon, and Martin G. Lockley. 1991. Dinosaur footprints with skin impressions from the Cretaceous of Alberta and Colorado. Canadian Journal of Earth Sciences, volume 28, pp. 102–115. (3) Lockley, Martin G. 1986. The paleobiological and paleoenvironmental importance of dinosaur footprints. Palaios, volume 1, pp. 37–47. (4) Lockley, Martin G. 1987. Dinosaur footprints from the Dakota Group of eastern Colorado. The Mountain Geologist, volume 24, number 4, pp. 107–122.

22. Florissant Fossil Beds National Monument

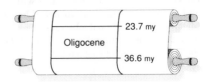

23.7 my

Oligocene

36.6 my

Florissant, Colorado

Florissant Fossil Beds National Monument is set in the Front Ranges of the Rocky Mountains, in a broad mountain valley on the eastern slopes of the continental divide at an elevation of about 2700 meters (9000 feet), with Pikes Peak to the east as a constant reference point. But 35 million years ago there was no Pikes Peak; the elevation, a mere 900 meters (3000 feet).

The high mountains of the early Tertiary Cordillera to the west had been subdued by Oligocene time. Prolonged erosion had carved and diminished the mountains; the sediment had been carried away and deposited in intermontane basins and on the Great Plains. The mountains were buried in their own debris, and a thick blanket of sediment extended southeastward across Wyoming, Colorado, Nebraska, and South Dakota. The best exposures of sediments contemporaneous with those at Florissant are in Badlands National Park [Site 17].

Intense volcanic activity during Oligocene time accounts for the spectacular preservation of fossils that characterizes Florissant Fossil Beds National Monument. Volcanoes in the Thirtyfive Mile Mountain Field began to erupt. Mudflows moved downslope burying trees and damming streams. An ephemeral lake, Lake Florissant, was created in this way, and fine-grained sediment and volcanic ash, ideal media for preserving delicate plants and insects, began to accumulate in it with successive volcanic eruptions.

The Tertiary rocks at Florissant Fossil Beds National Monument are volcanic in origin and include lava flows, ash deposits, river gravels, and lake shales. They can be subdivided into at least five distinct units, but only one, the lake shales, contains the delicate plant and insect fossils. Petrified stumps and trunks of the redwood, *Sequoia*, are common in the volcanic tuffs that underlie the lake shales. Two horizons are fossiliferous: the lower yields large trees, some up to 3 meters (10 feet) in diameter, in growth position; the upper contains smaller, uprooted trees oriented in a northerly and northeasterly direction. The rocks have been strongly faulted and eroded by subsequent geological activity so that the present distribution of rocks reflects the tectonic activity, not the original depositional setting.

The lake shales are famous for their plant and insect fossils. The fossils occur in fine-grained, fissile paper shales which are interbedded with volcanic ash and pumice. Several layers of paper shales are known, but fossils are most abundant in the lowest horizon. The preservation of the fossils, assured by rapid burial in an oxygen-poor setting, and the bedding of the sediments are attributed to alternating ponding and drying of the lake and episodic addition of very fine-grained, airborne volcanic ash.

The fossils in the paper shales are preserved as compressed organic remains and as impressions. They are the remains of the most delicate and fragile terrestrial organisms: insects, flowers, fruits, and leaves; organic remains usually quickly destroyed by decomposition and transportation. More than 100 species of plants alone are known. The Florissant fossil assemblage rates among the best three or four fossil insect faunas in the world. It includes bumblebees and wasps, many beetles and bugs, cockroaches, cicadas, caddisflies, termites, and many, many more. Even the ephemeral wings of a butterfly are exquisitely preserved. The fossils at Florissant Fossil Beds National Monument document an aspect of ancient life for which few good fossils are known.

The climate of Oligocene time can be inferred from the fossil plants. Annual mean temperatures were warm, but summers were hot. An average annual rainfall of approximately 50 centimeters (20 inches) was concentrated in the late spring and early summer. It was, therefore, only along streams and lakes, where moisture levels were higher, that true forests developed. A lush growth of giant redwoods, magnolias, sycamores, beeches, and oaks, some of them exotic trees foreign to the area today, hid an undergrowth of ferns and shrubs. On uplands and dry areas away from streams, scrub forest and grasses predominated.

Florissant Fossil Beds National Monument is dotted with petrified stumps of *Sequoia*. They are all that remain of a much larger forest of petrified trees long since destroyed by souvenir collectors. The shale beds are visible in the park, but the fossils themselves can only be seen by examining the large collection on display at the visitor center.

DIRECTIONS: Access to Florissant Fossil Beds National Monument is via United States Route 24 at Florissant. Follow County Route 1 south of United States Route 24 for 1 kilometer (0.6 mile) to the park boundary and continue on County Route 1 for another 3 kilometers (1.8 miles) to the visitor center access road (Figure 54). The visitor center is located 0.4 kilometer (0.25 mile) west of County Route 1.

PUBLIC USE: Season and hours: Florissant Fossil Beds National Monument is open year round: daily: from 8:00 A.M. to 4:30 P.M.; with extended hours Memorial Day to Labor Day to 7:00 P.M.; except closed New Year's Day (January 1), Thanksgiving Day, and Christmas Day (December 25). **Fees:** $3.00 daily or $25.00 annual Golden Eagle Permit. **Food service:** Restaurants and stores are available in Florissant. **Recreational activities:** Hiking, inter-

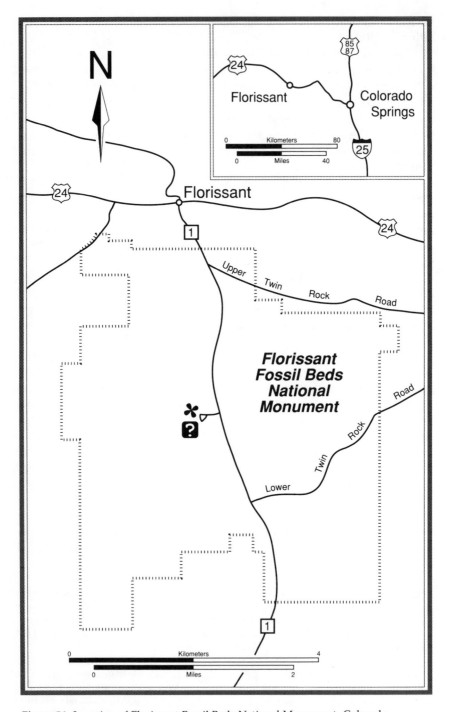

Figure 54. Location of Florissant Fossil Beds National Monument, Colorado.

preted natural and historical trails, and picnicking are available on site. **Handicapped facilities:** The visitor center and trails to some of the petrified trees are accessible by wheelchair. **Restrictions:** Collecting of fossils is prohibited.

EDUCATIONAL FACILITIES: Visitor Center: The visitor center houses an extensive collection of fossils recovered at the site. Slabs of the paper shale, split along bedding planes to exposed the fossils, are on display in cabinets behind glass. The nature of the fossils and the sediment in which they occur is so delicate that *in situ* exhibits are not feasible. Geological displays interpret the fortuitous events that led to the preservation of the fossils. Wildlife and human history are also featured. **Visitor Center hours:** The visitor center is open during park hours of operation. **Fees:** None. **Bookstore:** A small bookstore sells books of local interest as well as photographic slides of many of the fossils. **Trails:** Two self-guiding trails take visitors to the petrified trees that remain in the park. The Walk through Time Nature Trail is a short loop trail (0.8 kilometer or 0.5 mile) through the woods at the edge of the valley that features petrified and living trees. The Petrified Forest Loop Trail is a longer walk (1.6 kilometers or 1 mile) on the valley floor that winds past petrified trees and exposures of paper shales. Both trails are easy to walk. **Staff programs:** Park rangers offer a wide variety of interpretive programs during the summer. Of particular interest are the geological excursions that park rangers occasionally conduct locally and to the Garden Park Fossil Area, Cañon City, Colorado. Contact the visitor center for dates and times.

FOR ADDITIONAL INFORMATION: Contact: Superintendent, Florissant Fossil Beds National Monument, P. O. Box 185, Florissant, Colorado 80816. (719) 748–3253. **Read:** (1) Leopold, E. B., and H. D. MacGinitie. 1972. Development and affinities of Tertiary floras in the Rocky Mountains. Pp. 147–100 in Graham, A. (editor). Floristics and Paleofloristics of Asia and eastern North America. Amsterdam, Holland; New York, New York: Elsevier Publishing Company. (2) MacGinitie, H. D. 1953. Fossil Plants of the Florissant Beds, Colorado. Washington, D.C.: Carnegie Institution Publications, Number 599. (3) Saenger, Walter. 1982. Florissant Fossil Beds National Monument—Window to the Past. Estes Park, Colorado: Rocky Mountain Nature Association, Inc.

23. Garden Park Fossil Area

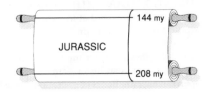

144 my

JURASSIC

208 my

Cañon City, Colorado

Deposits of the Morrison Formation form a highly varied but continuous blanket of sediment that was laid down in Late Jurassic time by rivers that flowed from highlands in the west. Water-borne debris was deposited as an alluvial plain on the margins of the epicontinental Sundance Sea; the sea was becoming filled with sediment and was gradually retreating northward. The rocks of the Morrison Formation are terrestrial in origin, having been deposited in river channels and floodplains, as overbank deposits, and as swamp and lake deposits.

Dinosaur bones were first discovered in the Garden Park Fossil Area in 1869 by M. P. Felch, a local rancher. In 1877, O. W. Lucas sent some fossil bones to Edward Drinker Cope and to Othniel Charles Marsh, adversaries in the infamous feud that colored both their careers and the early days of vertebrate paleontology in the United States. Both Cope and Marsh opened up dinosaur quarries, and the Garden Park area was quickly drawn into their feud. They excavated into stream channel and floodplain deposits, which proved to be exceptionally rich in dinosaur remains. Eight dinosaurs, each of them new to science at the time, were initially described; now fifteen are known from the area. Large sauropods are represented by *Apatosaurus*, *Diplodocus*, and *Camarasaurus*; theropod carnivores by *Allosaurus* and *Ceratosaurus*. Also present is the plated dinosaur, *Stegosaurus*, the state fossil of Colorado.

Fossils are not restricted to dinosaurs: fish, turtles, and crocodiles; fresh-water mollusks, arthropods, and ostracods; even primitive mammals are known. Nor have all the bones been removed. Old quarries and new exposures are being continuously eroded, and new fossils are exposed. The Garden Park Fossil Area figures large in the history of paleontology in North America and is the site of discovery (type site) of many species of dinosaurs. It also has tremendous potential for continued research and interpretive development, since it is clear that the rocks have not been exhausted of their dinosaur bones. Collecting of vertebrate fossils, however, is limited to scientific researchers. The Garden Park Fossil Area is undeveloped. Famous quarry sites, both old and

Figure 55. Dinosaur-bearing rocks of the Morrison Formation in Garden Park Fossil Area. The ancient stream channel within which a dinosaur skeleton was buried is clearly visible.

new, are visible, but seeing fossils is not guaranteed (Figure 55). The Garden Park Fossil Area has been designated a United States National Natural Landmark and is eligible for the National Register of Historic Places.

DIRECTIONS: Maps and directions to the Garden Park Fossil Area are readily available at the Cañon City District Office, Bureau of Land Management (Figure 56). The office is easy to find on East Main Street on the eastern outskirts of Cañon City as one approaches from the east on United States Route 50. Although access to the area is not restricted, the Bureau of Land Management is concerned about protecting the integrity of the sites and requests that visitors obtain information and directions before proceeding.

PUBLIC USE: Season and hours: The Garden Park Fossil Area comprises public land administered by the Bureau of Land Management and includes, in some places, adjacent private lands. Access to the area is encouraged by the Bureau of Land Management, and they will provide information on what is currently visible in exposure. In addition, the Florissant Fossil Beds National Monument (in adjacent Teller County) occasionally conducts interpretive tours to the Garden Park Fossil Area which are encouraged and supported by the Bureau of Land Management (schedule of events available at Florissant Fossil Beds National Monument, P. O. Box 185, Florissant, Colorado 80816). The Bureau of Land Management and the Garden Park Paleontology Society conduct their own tours all summer (information at the Cañon City Chamber of Commerce). The tours are recommended even for visitors with considerable geological and outdoor experience. The area is open year round, but weather conditions can limit access to and visibility of bone-bearing exposures of sediments. **Fees:** None. **Food service:** Restaurants and stores are available in Cañon City. **Recreational activities:** None on site. **Restrictions:** Collecting of fossils is prohibited.

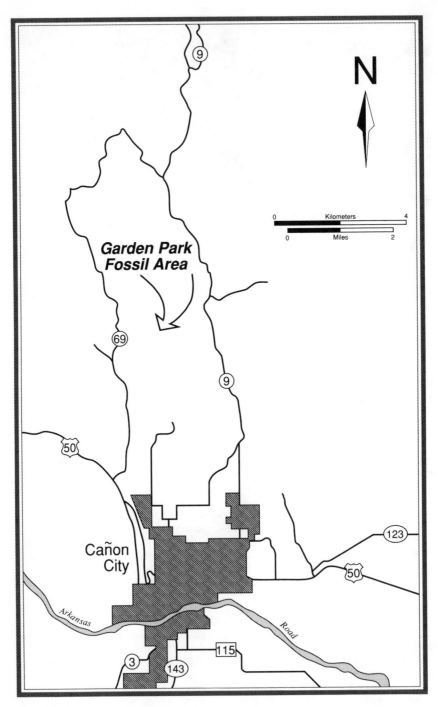

Figure 56. Location of Cañon City, Colorado, for access to Garden Park Fossil Area.

EDUCATIONAL FACILITIES: Interpretive sign: A roadside interpretive sign erected by the State Historical Society of Colorado to commemorate the discovery of dinosaurs here recognizes the historical significance of the area as the type locality for many species of dinosaurs. The sign is out-of-date and not entirely correct, but new interpretive signs are being developed and erected.

FOR ADDITIONAL INFORMATION: Contact: Bureau of Land Management, Cañon City District Office, 3170 East Main Street, P. O. Box 2200, Cañon City, Colorado 81215–2200, (719) 275–0631 *or* Garden Park Paleontology Society, P. O. Box 313, Cañon City, Colorado 81215–0313. **Read:** (1) Colbert, Edwin H. 1984. The Great Dinosaur Hunters and Their Discoveries. New York, New York: Dover Publications Inc. (2) Russell, Dale A. 1989. An Odyssey in Time: The Dinosaurs of North America. Toronto, Ontario: University of Toronto Press.

24. Indian Springs Trace Fossil Site

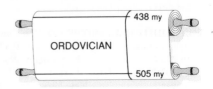

438 my

ORDOVICIAN

505 my

Cañon City, Colorado

During the Middle Ordovician, sea level was rising and epicontinental seas were once again flooding over the continent. Beaches were inundated, and river drainages converted to vast estuaries. In near-shore environments, sands and layers of clay were deposited; today they are known as the Harding Sandstone. Remains of the organisms that lived on the beaches and in the estuaries include trace fossils, shells, and the bony plates of primitive armored fish, *Astraspis*.

Approximately 60 million years had lapsed since the deposition of the Burgess Shale; during that time the first true vertebrates appeared. *Pikaia* from the Burgess Shale [Site 4], the first known chordate, establishes that the ancestors of vertebrates and the body plan which characterizes all vertebrates (Phylum Chordata) evolved during the early phases of the Cambrian Explosion. By Middle Ordovician time, 470 million years ago, vertebrates were apparently common, as their remains in the Harding Sandstone attest.

The fossils of the Harding Sandstone have resolved key questions about the evolution of the earliest vertebrates. The fish fossils, consistently associated with sandy deposits and not with the limestone deposits that formed in deeper water, show that the earliest vertebrates were indeed marine organisms living in the shallow, intertidal waters near the shores of epicontinental seas. The fish were jawless, obtaining their food by filter feeding; they were poor swimmers, covered in a layer of bony plates that afforded them some protection on the wave-agitated sea floor.

The exposure at Indian Springs Trace Fossil Site, designated a United States National Natural Landmark, consists of a pale, pink-buff mudstone bedding plane underlain and overlain by red sandstone. It is a horizontal surface formed of a thin layer of fine-grained mud that was deposited on top of sands and later covered by more sands. Where the upper sandstone has been removed, the tracks and traces of organisms that lived on the sea floor are easily seen (Figure 57). The trace fossils are more numerous and varied than are any other fossils at the site.

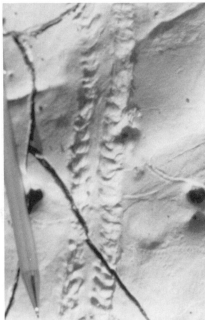
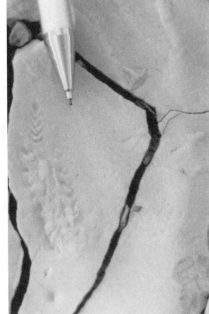

Figure 57. Outcrop and trace fossils at Indian Springs Trace Fossil Site. The upper photograph illustrates the mudstone bedding plane upon which trace fossils, such as those in the lower photographs, are readily visible. The pencil for scale is approximately 0.5 centimeter (0.2 inch) in diameter.

183

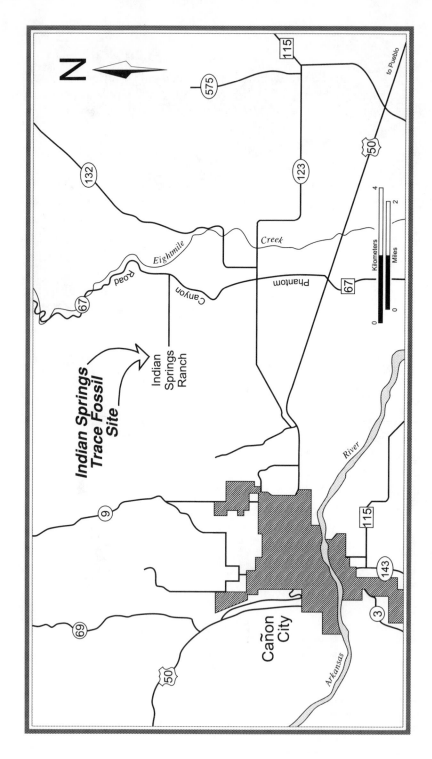

Figure 58. Location of Indian Springs Ranch near Cañon City, Colorado, for access to Indian Springs Trace Fossil Site.

It is one of the curious twists of history that Charles D. Walcott, years before his discovery of the Burgess Shale, had discovered and identified vertebrates from the Harding Sandstone. Just as he did not recognize the broad implications of the fauna from the Burgess Shale, Walcott did not recognize the presence of *Pikaia* among the fossils and its evolutionary connection to the Harding vertebrates. Nonetheless, his efforts at both localities, each housing critical pieces of evidence, contributed to the modern understanding of the early evolution of vertebrates.

DIRECTIONS: Follow United States Route 50 eastbound from Cañon City to the junction with State Route 67 (Phantom Canyon Road). Indian Springs Ranch entrance is located 6 kilometers (3.5 miles) north on State Route 67. The ranch house and campground is 2.2 kilometers (1.4 miles) west of State Route 67 (Figure 58).

PUBLIC USE: Season and hours: The Indian Springs Trace Fossil Site is located on Indian Springs Ranch, land privately owned by the Bennie Thorsen family. Access is by special permission of the land owners who escort visitors to the site. **Fees:** None. **Recreational activities:** Camping is available at Indian Springs Ranch (a fee is charged). **Food service:** Restaurants and stores are available at Cañon City. **Restrictions:** Collecting of fossils is prohibited.

EDUCATIONAL FACILITIES: Interpretive sign: A plaque commemorating the site as a United States National Natural Landmark is located on the site.

FOR ADDITIONAL INFORMATION: Contact: Mr. Bennie Thorsen, Indian Springs Ranch, Cañon City, Colorado 81212, (719) 275–2188 *or* Bureau of Land Management, Cañon City District Office, 3170 East Main Street, Cañon City, Colorado 81212, (719) 275–0613. **Read:** (1) Darby, David G. 1982. The early vertebrate, *Astraspis*, habitat based on lithologic association. Journal of Paleontology, volume 56, number 5, pp. 1187–1196. (2) Fischer, William A. 1978. The habitat of the early vertebrates: trace and body fossil evidence from the Harding Formation (Middle Ordovician), Colorado. The Mountain Geologist, volume 15, number 1, pp. 1–26.

25. Clayton Lake State Park

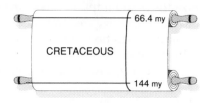

66.4 my

CRETACEOUS

144 my

Clayton, New Mexico

As Early Cretaceous time drew to a close, sea levels world-wide began to rise. Epicontinental seas encroached onto the western interior of North America; from the north the Beaufort Sea expanded southward, and from the south the Gulf Coast sea expanded northward (Figure 52). As the northern sea flooded farther southward and eastward across the interior of North America, sands were deposited on its shores, sands that now make up the Dakota Sandstone, a resistant layer of cap rock widely exposed in Colorado and New Mexico.

The tropical forests that flourished in the western interior of North America in the Early Cretaceous were home to a rich dinosaur fauna. Dinosaur bones are rare, however; the evidence of the animals is their footprints preserved in the Dakota Sandstone. The diversity of footprints attests to the numbers and variety of dinosaurs extant 100 million years ago.

At Clayton Lake State Park, dinosaur footprints occur along a bedding plane within the Dakota Sandstone. Four different kinds of dinosaur tracks are prominent. Most were made by heavy-bodied, bipedal dinosaurs such as duck-billed dinosaurs (hadrosaurs). The range in size of the footprints, the largest about 50 centimeters (20 inches) long by 42 centimeters (16 inches) wide, indicates that individuals of all ages, from juvenile to adult, were present. Almost as abundant are the tracks of various theropod dinosaurs: a small one is very bird-like, a larger one derives from a heavier animal, and an unusual one indicates a web-footed dinosaur. Other tracks are more difficult to interpret. Hundreds in one small area seem to have been made by birds or by small bipedal dinosaurs. A collection of tracks in a trackway suggests a pterosaur running to gain enough speed to begin flight, but the identity of the trackmaker remains equivocal because there is a suggestion that it could equally well indicate a crocodile.

Tracks offer unique insights in paleontology. They record the activity of the animals that made them and facilitate interpretation of dinosaur behavior. At Clayton Lake, some of the tracks are oriented in linear trackways indicating movement from one point to another; the majority, how-

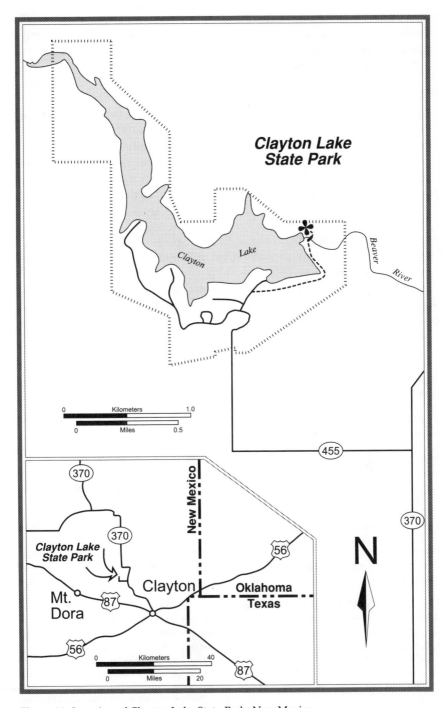

Figure 59. Location of Clayton Lake State Park, New Mexico.

ever, show no preferred orientation suggesting that the animals were simply milling about. Tracks also reveal the composition of dinosaur communities. Those at Clayton Lake and Dinosaur Ridge [Site 21] preserved in the Dakota Sandstone derive predominantly from ornithopod dinosaurs, large bipedal herbivores such as iguanodontids or hadrosaurs. The remaining one-third are attributable to theropods. Birds are represented by a third type of track, but they are a minute component of the track fauna. The tracks describe the dinosaur community that lived along the margins of the seaway advancing from the north.

The composition of the fauna that lived along the margins of the northern sea is in stark contrast to that of the coeval fauna living along the margins of the Gulf Coast sea as it advanced northward. At Dinosaur Valley State Park [Site 28], ornithopod tracks are a minor component (about 5 percent), more than two-thirds of the tracks are attributable to theropods and about one-quarter to sauropods. The tracks and their distribution are a clear indication of different ecological preferences among dinosaurs.

The tracks at Clayton Lake are located in the spillway of the dam. Approximately 500 dinosaur footprints are present. The fossils have not been damaged by erosion because the area has been dry since the dam was built.

DIRECTIONS: Clayton Lake State Park is 19.6 kilometers (12.2 miles) north of Clayton (Figure 59). Follow State Route 370 north from Clayton for 17.4 kilometers (10.8 miles) to State Route 455. The park is 2.2 kilometers (1.4 miles) northwest of the junction.

PUBLIC USE: Season and hours: Clayton Lake State Park is open year round. Access to the fossils is not limited, but the best time for viewing is early morning or late afternoon when the angle of the sunlight is low. **Fees:** $3.00 daily or annual New Mexico State Park Permit. **Food service:** Restaurants and stores available at Clayton. **Recreational activities:** Recreational activities such as swimming, picnicking, boating and fishing (March 1 to October 31), and camping (for which a fee is charged) are available. **Handicapped facilities:** The bath-house is accessible by wheelchair. **Restrictions:** Collecting of fossils is prohibited.

EDUCATIONAL FACILITIES: Trails: A pavilion, located at the level of the top of the dam and containing ten interpretive panels, overlooks the spillway. It provides access via a boardwalk through the fossil trackway site where a variety of tracks are randomly exposed. The short trail (approximately 1 kilometer or 0.5 mile) leads along the reservoir and across the dam.

FOR ADDITIONAL INFORMATION: Contact: Superintendent, Clayton Lake State Park, Seneca, New Mexico 88437, (505) 374–8808 *or* Clayton-Union County Chamber of Commerce, 1103 South First Street, Clayton, New Mexico 88415, (505) 374–9253. **Read:** (1) Gillette, David D., and Martin G. Lockley (editors). 1989. Dinosaur Tracks and Traces. Cambridge, England; New York, New York: Cambridge University Press. (2) Gillette, David D., and David A. Thomas. 1985. Dinosaur tracks in the Dakota Formation (Aptian-Albian) at Clayton Lake State Park, Union County, New Mexico. Pp. 283–288 in Lucas, Spencer G., and Jiri Zidek (editors). Santa Rosa Tucumcari Region, New Mexico Geological Society Guidebook, 36th Field Conference, Santa Rosa. (3) Lockley, Martin. 1986. The paleobiological and paleoenvironmental importance of dinosaur footprints. Palaios, volume 1, pp. 37–47.

26. Blackwater Draw

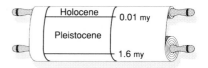

Portales, New Mexico

Late in the Pleistocene as the last ice sheet melted from the northern half of North America, big-game hunting people migrated from Asia across Beringia, the so-called Bering Land Bridge, into North America. They spread out relatively quickly across the continent, subsisting by hunting mammals now extinct: woolly mammoth, Columbian mammoth, and extinct species of bison. Other mammals that occupied the same environment are also extinct; among them camels, horses, saber-toothed cats, and dire wolves. Paleo-Indians lived at Blackwater Draw for several thousand years beginning about 11,000 years ago. A large pond along the drainage of the Brazos River was an ideal site for a hunting-gathering camp, for it would have attracted animals searching for water. The first human inhabitants hunted mammoths, and when the mammoths became extinct, the succeeding generations switched to hunting species of bison now extinct.

Blackwater Draw, historically also known as the Clovis Site, was discovered in 1932 by amateur collectors who found large projectile points and mammoth bones in association. These were new points, Clovis points, distinguished from Folsom points (Paleo-Indian points found in 1908 near Folsom in far northeastern New Mexico) by their much larger size and distinct shape. Subsequent archeological excavation revealed both types of points at Blackwater Draw: two layers of sediments contained bones of extinct species of bison and Folsom points in association; and beneath, a layer contained the mammoth bones and Clovis points. On top of the Folsom-bearing layer were artifacts associated with more recent cultures. The now-classical archeological sequence—Clovis-Folsom-Archaic—is represented at Blackwater Draw. A similar sequence is present at Lubbock Lake Landmark [Site 27].

Blackwater Draw established the antiquity of human occupation of North America and the association of immigrant peoples with Ice-Age mammals. For that reason it has been designated a United States National Historic Landmark. The site is undeveloped (although plans are being drawn for future development) and appears much as it did when the last

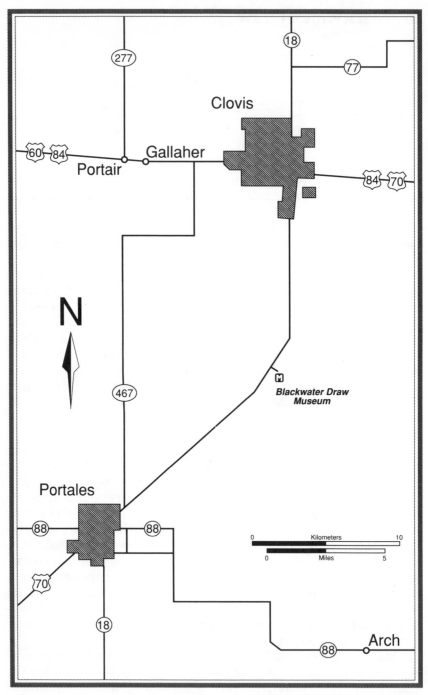

Figure 60. Location of Blackwater Draw Museum near Portales, New Mexico, for access to Blackwater Draw.

excavations were completed. In old excavation trenches one can see layers of bone, the remnants of successive butchering. The topography of the draw was strongly modified when the area was being quarried but enough of the original configuration remains to allow one to appreciate what the area was like when it was occupied by big-game hunters.

DIRECTIONS: Access to Blackwater Draw is by contacting the Blackwater Draw Museum located on United States Route 70 between Portales and Clovis, New Mexico (Figure 60): 9.5 kilometers (6 miles) from Portales; 16 kilometers (10 miles) from Clovis or the office of the Museum Director on the campus of Eastern New Mexico University in Portales.

PUBLIC USE: Season and hours: Public access is restricted to protect the integrity of the site, but visitors are encouraged and welcomed. Individuals, special interest groups, and schools can make arrangements to tour the site by contacting the Blackwater Draw Museum, a museum of Eastern New Mexico University or the Museum Director, Department of Anthropology, Eastern New Mexico University, Portales. **Fees:** None. **Food service:** Restaurants and stores are available in Portales and Clovis. **Recreational actiities:** None at the Blackwater Draw site or at the Blackwater Draw Museum. **Restrictions:** Collecting of fossils is prohibited.

EDUCATIONAL FACILITIES: Museum: Blackwater Draw Museum documents the archeological history of the area and the antiquity of human occupation. A plaque commemorating Blackwater Draw as a United States National Historic Landmark is located at the Museum. **Museum hours:** Museum hours are highly variable. Visitors should seek additional information when planning a visit. **Special group activities:** Special activities for interested groups (school groups, visiting groups, etc.) can be arranged by contacting the Museum Director, Department of Anthropology, Eastern New Mexico University, Portales.

FOR ADDITIONAL INFORMATION: Contact: Museum Director, Department of Anthropology, Eastern New Mexico University, Portales, New Mexico 88130, (505) 562–2254. **Read:** (1) Agenbroad, Larry D., Jim I. Mead, and Lisa W. Nelson (editors). 1990. Megafauna and Man: Discovery of America's Heartland. Hot Springs, South Dakota: The Mammoth Site of Hot Springs, South Dakota, Inc. Scientific Papers, Volume 1. (2) Fagan, Brian M. 1987. The Great Journey: The Peopling of Ancient America. London, England; New York, New York: Thames and Hudson Ltd. (3) Klein, Richard G., and Kathryn Cruz-Uribe. 1984. The Analysis of Animal Bones from Archeological Sites. Chicago, Illinois: The University of Chicago Press.

27. Lubbock Lake Landmark

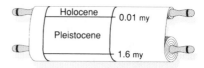

Lubbock, Texas

Human populations migrated across Beringia and into North America in the Late Pleistocene and spread rapidly southward and eastward across the continent as the last ice sheet melted away. These people were big-game hunters, pursuing first the mammoths and then the bison. Human-habitation sites dating from this time, 12,000 or 11,000 years ago, are characterized by the unique association of human artifacts and the bones of extinct animals. The oldest artifacts are the large, so-called Clovis points; at some localities, Clovis points are found embedded in mammoth bones (as they were at Blackwater Draw [Site 26]). When the mammoths became extinct, the big game hunters turned to bison. Bison-kill sites are younger than mammoth sites and contain the distinctive Folsom point. Among the most southern of the human-habitation sites in North America is Lubbock Lake Landmark.

The Lubbock Lake Landmark is on the site of a Late Pleistocene draw that began to fill with sediments about 12,000 years ago. The oldest deposits, sands and gravels deposited by a stream, contain human artifacts and the bones of extinct large mammals. Eventually a pond developed, and the sediments that accumulated on the edge of the pond contain Folsom points. Younger sediments contain successively younger habitation sites; the most frequently encountered evidence for human occupation is charcoal in firepits and hearths. The draw offered shelter and water for the nomadic hunters; for almost 3000 years, as deposition in the draw continued, human populations regularly, if not continuously, inhabited the area. The Lubbock Lake Landmark, a United States National Historic Landmark, and nearby Blackwater Draw [Site 26] confirm the coexistence of Paleo-Indians and extinct mammals. Lubbock Lake Landmark offers a window on the world of the first people in North America, and for that reason, it has recently been developed as a major scientific research center. Equally important is the emphasis being placed on developing the site as an interpretive center for public access; the new facility opened to the public in October, 1990.

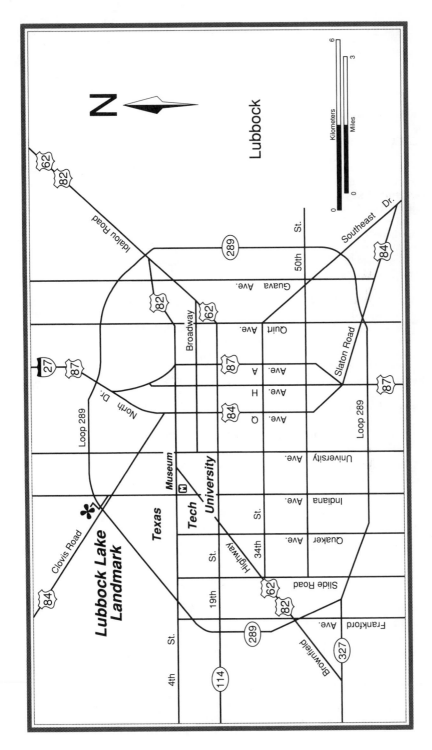

Figure 61. Location of the Lubbock Lake Landmark, Lubbock, Texas.

DIRECTIONS: Lubbock Lake Landmark is located within the city of Lubbock, north of Loop 289 and Clovis Road (United States Route 84) on the northwest edge of the city (Figure 61). Access is by way of Indiana Avenue.

PUBLIC USE: Season and hours: Lubbock Lake Landmark is open year round: daily; from 8:00 A.M. to 5:00 P.M.; with extended hours May to September to 10:00 P.M. **Fees:** None. **Food service:** Restaurants and stores are available in Lubbock. **Recreational activities:** Picnicking is available on site. **Restrictions:** Collecting of fossils is prohibited.

EDUCATIONAL FACILITIES: Visitor Center: The Robert A. "Bob" Nash Interpretive Center offers a full range of educational facilities such as interpretive programming and a hands-on learning center. The Museum of Texas Tech University is also suggested for visitors interested in the history of human occupation and archeology in the area. **Visitor Center hours:** The visitor center is open during the Lubbock Lake Landmark hours of operation. **Fees:** None. **Bookstore:** A giftshop which stocks books on topics of local interest is housed in the visitor center. **Trails:** Short, easy-to-walk interpretive trails have been developed, and guided tours of the site are given daily. **Note:** The Lubbock Lake Landmark and the Museum of Texas Tech University cooperatively offer a Texas Paleohistory and Paleoecology Field Program, a summer field season for volunteers and for which university credit can be arranged.

FOR ADDITIONAL INFORMATION: Contact: Director, Lubbock Lake Landmark, The Museum of Texas Tech University, Fourth and Indiana Avenues, Lubbock, Texas 79409, (806) 742–2481. **Read:** (1) Agenbroad, Larry D., Jim I. Mead, and Lisa W. Nelson (editors). 1990. Megafauna and Man: Discovery of America's Heartland. Hot Springs, South Dakota: The Mammoth Site of Hot Springs, South Dakota, Inc. Scientific Papers, Volume 1. (2) Fagan, Brian M. 1987. The Great Journey: The Peopling of Ancient America. London, England; New York, New York: Thames and Hudson Ltd. (3) Klein, Richard G., and Kathryn Cruz-Uribe. 1984. The Analysis of Animal Bones from Archeological Sites. Chicago, Illinois: The University of Chicago Press.

28. Dinosaur Valley State Park

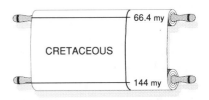

CRETACEOUS

66.4 my

144 my

Glen Rose, Texas

The coastal plain of the Gulf of Mexico extended much farther inland in Early Cretaceous time than it does today, and as sea level rose, the water began to flood northward onto the western interior of North America. An elongate barrier reef, thousands of kilometers long, grew along the margin of the continent; inland, vast lagoonal flats and mangrove-like swamps formed on the margin of the shallow epicontinental sea (Figure 52). The sea was clear and brimming with life. Calcareous debris and lime muds accumulated in the sea and on the shore.

The Glen Rose Formation, which contains many trackways of dinosaurs, crops out over a large area of central Texas. The rocks are shallow-water limestones that formed about 105 million years ago in lagoons on the landward side of the barrier reef. The lagoons, protected though they were from the high energy of the open ocean, were nonetheless subject to daily intertidal wave energy and to supratidal storm action. The limestone rocks are of variable composition, are altered to dolomite in many places, and are regularly interbedded with shales. The latter characteristic is significant for the preservation of dinosaur tracks. There are at least two track-bearing layers at Dinosaur Valley State Park. In each, the tracks themselves occur in the limestones and are infilled with shales. The preservation of the tracks, made as animals walked along the edge of the lagoon during low tide, was assured by subsequent high water that flooded over the tracks; fine mud settled out of the water as the flood receded and buried the tracks.

The trackways are attributable to sauropod and moderate-sized theropod dinosaurs. The exact species of dinosaur that made the tracks cannot be identified, but a wealth of information about the behavior of the track makers can be derived. The size and weight of the animal, the speed and cadence with which the animal moved, and swimming behavior can be interpreted. Associated groups of herbivore trackways reflect herd behavior; solitary carnivore trackways, their solitary lifestyle. Parallel trackways probably mark the ancient coastline. Tracks of the same kind of animal going in opposite directions may indicate daily movement to and from feeding areas or watering holes. Even environmental parameters such as the depth of the water can be calculated.

The dinosaur fauna that lived along the lagoon and swamps, and in the coastal-plain habitats of the expanding Gulf Coast sea was dominated by theropods, represented by more than two-thirds of the tracks, and sauropods. Ornithopod tracks form only about 5 percent of the fauna. This community of dinosaurs is distinctly different from the dinosaur fauna that existed contemporaneously in the western interior of North America on the shores of the encroaching northern sea. At Clayton Lake State Park [Site 25], for example, about two-thirds of the tracks are attributed to ornithopods, most of the rest to theropods. The faunal differences reflect the unique ecological requirements of different communities of dinosaurs.

Dinosaur tracksites in Texas have long been known and are famous. The first museum collectors quickly recognized that the variety of footprints offers insight into dinosaur behavior. Among the trackways removed from the current site of Dinosaur Valley State Park for display at the American Museum of Natural History and at Texas Memorial Museum by the Bird expedition in 1939 were parallel trackways of a sauropod and a carnosaur, interpreted then as a carnivore pursuing its prey. The significance of the trackways is underscored by the designation of Dinosaur Valley State Park as a United States National Natural Landmark.

Dinosaur trackways and footprints, exposed in the bed of the Paluxy River and tributary creeks, are visible on the stream bottom in several places within the park. When water levels are low it is possible to walk on the track-bearing layers of limestone and to follow the layers where they extend beneath overlying rocks on the banks of the river.

DIRECTIONS: The junction of Farm Road 205 with United States Route 67 is on the western outskirts of the town of Glen Rose. Follow Farm Road 205 north for 4.8 kilometers (3 miles) to its junction with Park Road 59; proceed on Park Road 59 for 1 kilometer (0.6 mile) to the park office and visitor center (Figure 62).

PUBLIC USE: Season and hours: Dinosaur Valley State Park is open year round. **Fees:** $2.00/vehicle daily or annual Texas State Park Permit. **Food service:** Restaurants and stores are available in Glen Rose. **Recreational activities:** Recreational activities include hiking, fishing, swimming, and camping (including primitive camping) for which an additional fee is charged. **Restrictions:** Collecting of fossils and casting of trackways is prohibited.

EDUCATIONAL FACILITIES: Visitor Center: The visitor center, located near the park entrance, features exhibits and audiovisual displays which analyze dinosaur trackways and interpret behavior. An excellent trackway is on display outside. **Visitor Center hours:** The visitor center is open year round: daily; from 8:00 A.M. to 5:00 P.M. **Fees:** None. **Trails:** Dinosaur trackways are easily seen from two locations within the park. Access to each is along a park road to a parking lot on the river bank. One can view the fossils from the bank or walk down to the river bed for a closer view. A park map detailing the location of track sites is available at the visitor center.

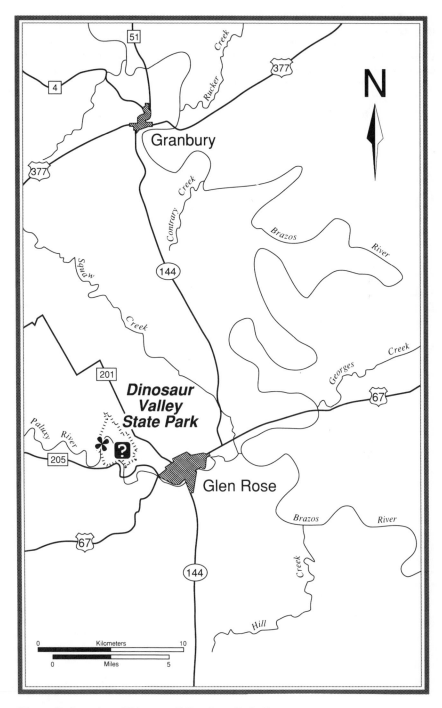

Figure 62. Location of Dinosaur Valley State Park, Texas.

FOR ADDITIONAL INFORMATION: Contact: Superintendent, Dinosaur Valley State Park, Glen Rose, Texas 76043, (817) 897–4588. **Read:** (1) Bird, Roland T. 1985. Bones for Barnum Brown: Adventures of a Dinosaur Hunter. Fort Worth, Texas: Texas Christian University Press. (2) Farlow, James O. 1987. A Guide to Lower Cretaceous Dinosaur Footprints and Tracksites of Paluxy River Valley, Somerville County, Texas. Geological Society of America, South Central District, Baylor University. (3) Gillette, David D., and Martin G. Lockley (editors). 1989. Dinosaur Tracks and Traces. Cambridge, England; New York, New York: Cambridge University Press.

29. Big Bend National Park

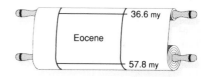

36.6 my

Eocene

57.8 my

Marathon, Texas

The geology of Big Bend National Park is complex, the scenery unique, and the park is justifiably famous for its geological attractions. Among the wide variety of rocks exposed within the park are many sedimentary rocks which contain a surprising range and wealth of fossils including invertebrate fossils, dinosaurs, and mammals. The three distinct kinds of fossils represent distinct times in the geological history of the earth. One group comprises marine organisms of Paleozoic age deposited when shallow seas covered the area; another is made up of marine invertebrates and terrestrial reptiles, including dinosaurs, of Cretaceous age; the third, Tertiary mammals. Park interpretation, however, focuses on the latter; in particular, the mammals of Eocene age.

Both the geological history and the fossil record of Big Bend National Park are directly linked to the tectonic events that assembled the supercontinent of Pangea and, in the process, built the Ouachita Mountains. The rocks exposed in the park document a long history that dates to Late Precambrian time.

In Late Precambrian and Early Cambrian time, North America was sutured along its southern margin to a proto-South American continent; but the two landmasses were rifted apart in Early Paleozoic time, and an ocean formed between them. Among the sediments that accumulated in that ocean were limestones that contain brachiopods and trilobites. The exposure of these fossiliferous rocks is limited to those areas such as the Marathon Uplift that have been exposed by recent uplift and erosion; elsewhere they remain deeply buried. Late in the Paleozoic Era, Gondwana, the newly-assembled giant continent of the southern hemisphere, drifted northward and collided with eastern and southern North America, at that time, part of the large northern continent, Laurasia. Mountains were built as the continents of the world collided to form Pangea.

Rifting began again early in the Mesozoic Era (Triassic), Pangea was sundered, and deep ocean basins developed between the landmasses.

Throughout the Mesozoic, sediments accumulated in the basins, and as sea levels fluctuated, the deep ocean waters spilled over on the continent in the form of epicontinental seas. Shallow-water seas encroached on the land in the Late Cretaceous, a long barrier reef developed off-shore, and an extensive layer of limestone was deposited. Near shore, where lagoons and swamps were prevalent and dinosaurs were common, shallow-water lime muds were deposited to form the Glen Rose Formation (see Dinosaur Valley State Park, [Site 28]). Farther from shore, the area that is now Big Bend National Park was covered in deeper water; the Santa Elena Formation was deposited, richly fossiliferous with typical Late Cretaceous organisms such as *Inoceramus*, the giant clam. Later, about 75 million years ago, the area became subaerially exposed, extensive forests developed, and dinosaurs thrived. Petrified wood and dinosaur bones are found in the Aguja and Javelina formations. Contemporary deposits such as those at Dinosaur Provincial Park [Site 2] and Willow Creek Anticline [Site 12] show how diverse the ecological requirements of dinosaurs were.

The area of Big Bend National Park was subaerially exposed for most of Tertiary time. The sediments that were deposited in the lakes and rivers preserve an array of mammal fossils. At the Fossil Bone Exhibit, mammals of Eocene age are featured, but the panorama of fossiliferous rocks spans 20 million years of earth history. *Phenacodus*, a member of the archaic group from which all odd-toed ungulates are descended; *Hyracotherium*, the earliest known ancestor of modern horses; and the tapir-like, hoofed herbivore, *Coryphodon*, are exhibited. That these mammals were widely distributed across western North America is indicated by similar faunas at John Day Fossil Beds National Monument [Site 7] and Fossil Butte National Monument [Site 15].

Fossiliferous rocks are plentiful in Big Bend National Park: invertebrate fossils are readily visible wherever marine limestone are exposed; by comparison, vertebrate fossils are rare and unevenly distributed.

DIRECTIONS: Access to Big Bend National Park is via United States Route 385 from Marathon, a distance of 110 kilometers (69 miles), via State Route 118 from Alpine 166 kilometers (103 miles), or via United States Route 67 from Marfa (through Presidio) 246 kilometers (153 miles) to park headquarters at Panther Junction (Figure 63). The park occupies a large geographical area in the Big Bend of the Rio Grande; only the main roads are paved, and distances within the park are great. For example, the campground nearest to Panther Junction is The Basin, 16 kilometers (10 miles) away.

PUBLIC USE: Season and hours: Big Bend National Park is open year round, but visitor services and facilities are limited in the summer off-season (Memorial Day to Labor Day). Access to fossils is not restricted although only one spot is developed for ready public access. Visitors, particularly hikers, who are willing to explore will see many fossils *in situ*.
Fees: $5.00/vehicle or $2.00/person (entering by commercial bus, bicycle, or motorcycle) 7-day permit or $25.00 annual Golden Eagle Permit (inquire about other park permits).
Food service: Stores and a restaurant operated by a concessioner are available in the park.
Recreational activities: A wide variety of park facilities and outdoor activities are available

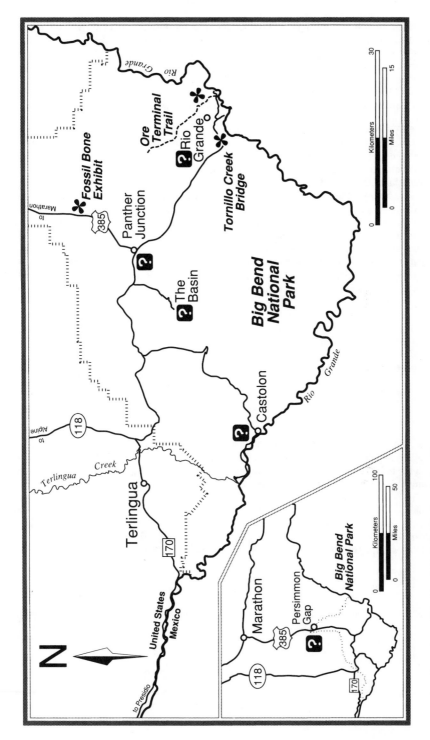

Figure 63. Location of Big Bend National Park, western Texas.

including park interpretive programs, camping (a fee is charged in developed campgrounds, and a free permit is required to camp in the back country), hiking, backpacking, horseback riding, river excursions, etc. Note that Big Bend National Park is isolated from urban areas, and the nearest town is more than 100 kilometers (60 miles) away. As a result, services such as gasoline stations and general stores are available in the park, but availability of goods may be limited. **Handicapped facilities:** The visitor center and Fossil Bone Exhibit are accessible by wheelchair. **Restrictions:** Collecting of fossils is prohibited.

EDUCATIONAL FACILITIES: Visitor Center: The park headquarters and main visitor center is at Panther Junction. It has extensive displays and interpretive information, with a newly developed emphasis on the fossils found in the park. Four other centers, with limited display material, serve the more distant areas of the park. **Visitor Center hours:** Panther Junction Visitor Center is open year round: daily; from 8:00 A.M. to 6:00 P.M.; with extended hours in the cooler seasons. The other visitor centers have shorter hours that are variable with visitor demand and availability of staff. **Fees:** None. **Bookstore:** The bookstores at the visitor centers are well-stocked to meet the needs of park visitors, especially important because the park is remote. Maps, guides, and books on fossils, geology, and natural and human history are all available. **Trails:** Fossils in Big Bend National Park can readily be seen in one of two ways. The Fossil Bone Exhibit, located 14 kilometers (8.5 miles) north of Panther Junction on United States Route 385, at the end of a very short trail illustrates fossils in an interpreted setting. On the other hand, invertebrate fossils can be seen along many of the desert trails and canyons in the low country wherever limestone rocks are exposed. For example, limestone rocks are exposed at the bridge on Lower Tornillo Creek, and large Cretaceous marine fossils like *Inoceramus* can be seen; and near the trail head of Ore Terminal Trail many fossils can be seen on the floor of the small wash which the trail traverses. There are many trails in the park ranging from self-guiding, interpreted trails to primitive routes, but none feature fossils. This is particularly true of the most popular trails, those in the High Chisos Mountains which traverse volcanic terrain. **Staff programs:** Naturalist activities such as interpretive programs and ranger-led hikes are available especially during peak visitor season from November through April (schedules are posted).

FOR ADDITIONAL INFORMATION: Contact: Superintendent, Big Bend National Park, Texas 79834, (915) 477–2251. **Read:** (1) Maxwell, Ross A. 1968. Big Bend of the Rio Grande: A Guide to the Rocks, Landscape, Geologic History, and Settlers of the Area of Big Bend National Park. Bureau of Economic Geology, The University of Texas at Austin, Guidebook F. (2) Maxwell, Ross A., John T. Lonsdale, Roy T. Hazzard, and John A. Wilson. 1967. Geology of Big Bend National Park, Brewster County, Texas. Austin, Texas: University of Texas, Bureau of Economic Geology Publication, Number 6711. (3) Pausé, P. H., and R. Gay Spears (editors). 1986. Geology of the Big Bend Area and Solitario Dome, Texas. West Texas Geological Society 1986 Field Trip Guidebook, Publication 86–82.

30. Capitan Reef
Guadalupe Mountains National Park

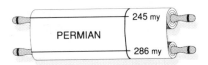

245 my

PERMIAN

286 my

Pine Springs, Texas

Two hundred and fifty million years have passed since the Capitan Reef in western Texas was a living community of organisms. Today, in spite of the intervening time and inexorable geological processes, the reef stands high on the margins of an inland basin, encircling the basin as it did during Late Permian time. The modern basin is a parched desert; in the Permian it was a deep-water basin that supported a myriad of plant and animal life along its margins.

During Permian time, the coalescing land masses of the earth were being finally assembled into Pangea, a giant supercontinent straddling the equator. As land masses collided, huge mountain ranges were built along the lines of contact, and deep ocean basins developed adjacent to the mountains. The Ouachita Mountains formed then, high ranges the remains of which now bisect Texas along a northeast-southwest line from Oklahoma to Mexico. To the north and west, the continent was extensively covered by a shallow sea (Figure 64). The mountains were elevated; the basins, in particular the Delaware Basin, became deeper; and a pronounced rainshadow developed in the lee of the mountains. Organic growth was prolific along the margins of the Delaware Basin; limestones to a thickness of 600 meters (2000 feet) accumulated in a giant horseshoe rimming the basin (Figure 65). Algae, sponges, and bryozoans; rugose corals, brachiopods, and echinoderms secreted the calcium carbonate that now forms El Capitan and the Capitan Limestone. Within the basin itself, the water was deep (almost 2000 meters or 6500 feet); the nutrient supplies were lower, oxygen was depleted, and few organisms lived. Eventually the growth of the reef restricted the circulation of marine water from the open ocean to the southwest into the Delaware Basin. Such restriction, when combined with the evaporation of an arid climate, led to the deposition of a thick layer of evaporite carbonates and salts within the basin. Behind the reef in shallow epicontinental water, the fine debris of broken shells was washed and winnowed, lime muds deposited as the Carlsbad Limestone. Evaporite rocks filled in the basin,

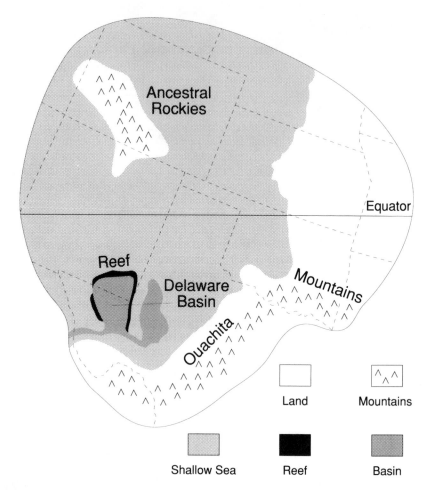

Figure 64. Paleogeography of the southern margin of proto-North America during Late Permian time. The Ouachita Mountains had been built as Pangea was assembled.

buried both the reef and the lime muds behind it, and provided a protective cap over the entire region.

Today the Capitan Reef, a 250 million years old biological community frozen in time, forms the Guadalupe Mountains. It has been uplifted by tectonic processes and exhumed by erosion, its three-dimensional structure revealed because the evaporite rocks which once covered the entire area are much more readily removed than are the limestones that formed on the margins of the basin. El Capitan is the visible corner stone of the Capitan Reef (Figure 66). Hiking the Permian Reef Trail takes one inside the reef and provides a unique picture of ancient paleogeography: the basin, the reef, the epicontinental sea bed, all the physical features from the Permian remain; all that is missing is the water.

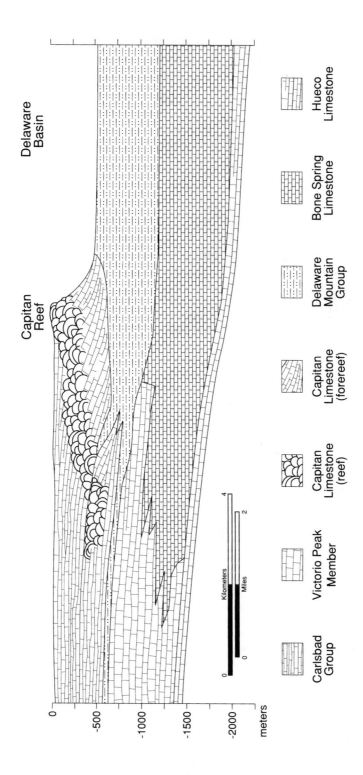

Figure 65. Cross-section through the Capitan Reef.

205

Figure 66. El Capitan, Guadalupe Mountains National Park.

The irony of the Capitan Reef is that it was not a reef in any true biological sense. It was built, not by reef-building organisms—not by organisms that build a three-dimensional structure on the sea floor as they secrete their skeletons of calcium carbonate—but by fragile organisms that lived in deeper, quieter water below the margin of the basin. The biota was able to simulate a reef because it grew prolifically along the margins of a rapidly subsiding basin where the water was rich in nutrients, the unique setting allowing a great thickness of reef-like limestone to accumulate. The present relief of the limestone formations, that which allows the limestone to be called a reef, is an artifact of differential erosion; it was not characteristic of the living community.

DIRECTIONS: Guadalupe Mountains National Park can be reached from the south via State Route 54 from Van Horn to the junction of United States Route 62/180, a distance of 94 kilometers (58 miles). The park boundary is 13 kilometers (8 miles) north of the junction, and the Pine Springs Visitor Center is 2 kilometers (1.3 miles) beyond that (Figure 67). From the northeast, the park is 56 kilometers (35 miles) from Whites City on United States Route 62/180. There is also access to the north side of the park via El Paso Gap on State Route 137.

PUBLIC USE: Season and hours: Guadelupe Mountains National Park is open year round although access to McKittrick Canyon and to two geological trails is restricted to daylight hours. **Fees:** None. **Food service:** Nearest food services are at Whites City. **Recreational activities:** The Park has 130 kilometers (80 miles) of hiking trails ranging from well-developed to primitive. Horses are welcome on many of them. There are two well-developed

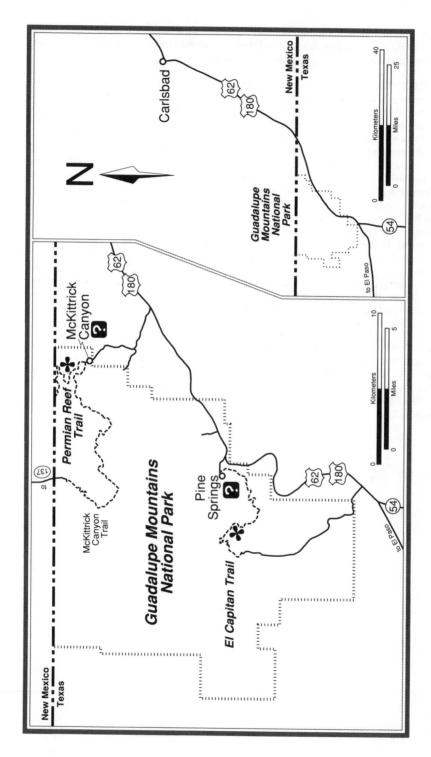

Figure 67. Location of Guadalupe Mountains National Park, western Texas.

campgrounds (a fee is charged at one), and permits for back-country camping in designated areas can be obtained. **Handicapped facilities:** The visitor centers are accessible by wheelchair. **Restrictions:** Collecting of fossils is prohibited.

EDUCATIONAL FACILITIES: Visitor Center: There are two visitor centers in the park. The Pine Springs Visitor Center also functions as park headquarters. It is the main center of information about the park, and it exhibits fossils and rocks from the reef. The McKittrick Canyon Visitor Center is an outdoor center with interpretive panels and audio-visual presentations highlighting the natural history of the area, in particular McKittrick Canyon. **Visitor Center hours:** The Pine Springs Visitor Center is open year round: daily; from 8:00 A.M. to 4:30 P.M.; with extended hours in summer from 7:00 A.M. to 6:00 P.M. The McKittrick Canyon Visitor Center is open for the visitor season during daylight hours. **Fees:** None. **Bookstore:** A small bookstore featuring books and maps is present in both visitor centers. **Trails:** The rocks in the park are predominantly limestones and contain shelly fossils. Two trails focus directly on geology and paleontology. One is a short, self-guided walk (0.7 kilometer or 0.4 mile) at the McKittrick Canyon Visitor Center which highlights geological features of the park. The most impressive features of the area, however, will best be seen by actually hiking into the mountains. One can get a feel for the magnitude of the ancient reef by hiking the Permian Reef Trail, a strenuous 7.5 kilometer (4.6 mile) one-way hike along which you actually climb the reef and then walk along it. It is worth noting that any excursion in the Guadalupe Mountains is a hike along the reef. **Staff programs:** Ranger naturalist programs, including guided hikes and evening talks, are available mid-May through mid-September.

FOR ADDITIONAL INFORMATION: Contact: Superintendent, Guadalupe Mountains National Park, HC 60, Box 400, Salt Flat, Texas 79847–9400, (915) 828–3251. **Read:** Newell, Norman D., J. Keith Rigby, Alfred G. Fischer, A. J. Whiteman, John E. Hickox, and John S. Bradley. 1953. The Permian Reef Complex of the Guadalupe Mountains Region, Texas and New Mexico. A Study in Paleoecology. San Francisco, California: W. H. Freeman and Company.

31. Ghost Ranch *Coelophysis* Quarry
Ghost Ranch Conference Center

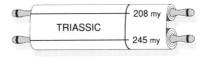

TRIASSIC
208 my
245 my

Abiquiu, New Mexico

Ghost Ranch is the home of *Coelophysis*, one of oldest and most primitive carnivorous dinosaurs known. *Coelophysis*, despite its antiquity, is clearly a dinosaur whose features foreshadow famous carnivores, such as *Allosaurus* and *Tyrannosaurus*. It was an agile predator with bipedal posture and a long, muscular, counter-weight tail. It had short forelimbs with elongate, sharp, recurved claws; a large mouth with numerous, sharp, knife-like teeth. Dinosaurs are known from other Triassic localities in North America, some are found in nearby Petrified Forest National Park [Site 32], but none are as complete or as numerous as *Coelophysis* from Ghost Ranch.

The Four Corners region of the United States, 225 million years ago, was a flat, lowland plain surrounded by discontinuous highlands (Figure 68). The plain was drained by rivers that flowed northwestward into the ancient Pacific Ocean, toward a shoreline which lay in central Nevada. The lowland was heavily forested, its waterways providing habitat for a variety of animal life that included amphibians and alligator-like reptiles, both carnivores and herbivores. In the drier, upland regions surrounding the plain, early dinosaurs evolved. In the open and often sparse shrubbery that predominated, the agility, speed, and keen eyesight of *Coelophysis* were valuable characteristics. Other dinosaurs must have been present there, too. In nature's scheme of things, however, upland areas tend to be places of erosion rather than deposition, and the organisms that lived there are poorly represented in the geological record.

Only fortuitous circumstance accounts for the remarkable evidence of *Coelophysis* preserved at Ghost Ranch. The *Coelophysis* Quarry at Ghost Ranch occurs in the Petrified Forest Member of the Chinle Formation. The strata, equivalent to those exposed at Petrified Forest National Park, were deposited by rivers that flowed across the tropical plain. The quarry contains at least 100 complete skeletons of individuals that died in and were buried by a flash flood, a flood that deposited carcasses in an area significantly different than the one in which the animals lived.

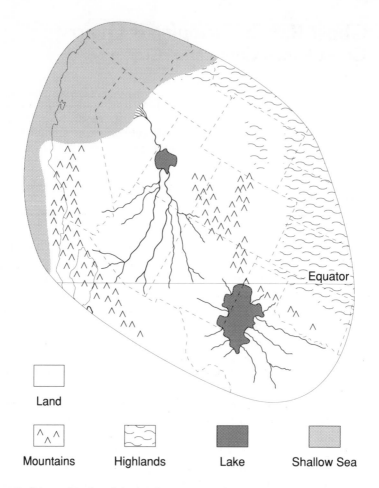

Figure 68. Paleogeography of the southern margin of North America during Late Triassic time.

Ghost Ranch *Coelophysis* Quarry, a United States National Natural Landmark, has been closed to fossil collecting for the present time and the quarry face has been covered with sediment to protect it from erosion. The area, however, beautifully displays the spectacular and characteristic outcropping of the Chinle Formation. Fossils do weather out of outcrop from time to time and may well be seen during a visit.

DIRECTIONS: Ghost Ranch *Coelophysis* Quarry located at the Ghost Ranch Conference Center on United States Route 84 is 24 kilometers (15 miles) northwest of Abiquiu in Rio Arriba County, New Mexico (Figure 69).

PUBLIC USE: Season and hours: Ghost Ranch *Coelophysis* Quarry is privately owned by the Ghost Ranch Conference Center. It is open to the public as well as to formally organized conferences and functions. The public is directed to the Ruth Hall Museum of Paleontology

210

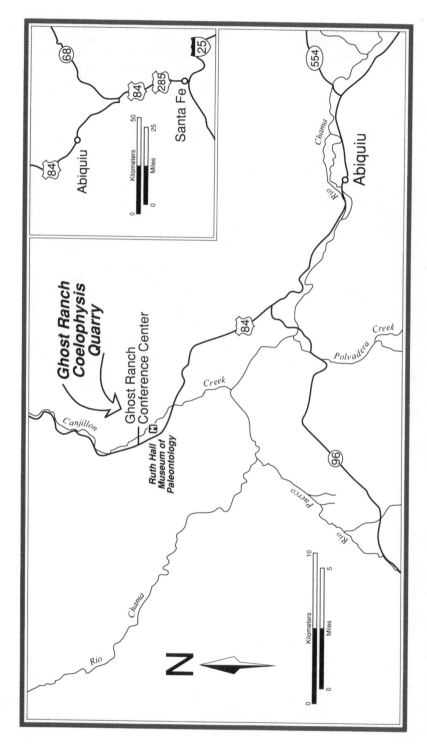

Figure 69. Location of the Ghost Ranch Conference Center near Abiquiu, New Mexico, for access to Ghost Ranch *Coelophysis* Quarry.

where guided tours to the fossil site can be arranged. **Fees:** None. **Food service:** The conference center dining hall is open to the public, but advance notice should be given at the office. **Recreational activities:** Ghost Ranch is open to the public for camping (for which a fee is charged) and has self-guided hiking trails. Since it is not primarily a tourist area, facilities may be limited or unavailable if the center is hosting a conference. **Handicapped facilities:** Public use buildings are accessible by wheelchair. **Restrictions:** Collecting of fossils is prohibited.

EDUCATIONAL FACILITIES: Museum: Ruth Hall Museum of Paleontology is the focus of educational programming in paleontology at Ghost Ranch. Featured is a *Coelophysis*-bearing block, which was removed from the quarry and is housed in the museum. Here a museum preparator will work on the material during museum hours and visitors are invited to watch. Archeology is featured in a separate museum. Some educational and interpretive programming is available, and arrangements for groups and organizations can be made. **Museum hours:** The museum is open year round: daily in summer; Tuesday to Saturday from 9:00 A.M. to 12:00 A.M. and 1:00 P.M. to 5:00 P.M., Sunday and Monday 1:00 P.M. to 5:00 P.M.; Saturday and Sunday in winter about October 15 to May 15; from 1:00 P.M. to 5:00 P.M. **Fees:** None. **Bookstore:** The Ghost Ranch Conference Center has a bookstore which carries some titles in natural history. **Special group activities:** Special group activities can be arranged by contacting the Ruth Hall Museum of Paleontology. **Note:** In the summer, Ghost Ranch Conference Center runs a program of week-long seminars, among them a seminar on the paleontology of northern New Mexico. In the spring and fall, Elderhostel classes in paleontology are also offered at the ranch.

FOR ADDITIONAL INFORMATION: Contact: Curator, Ruth Hall Museum of Paleontology, Ghost Ranch Conference Center, Abiquiu, New Mexico 87510, (505) 685–4333. **Read:** (1) Colbert, Edwin H. 1989. The Triassic Dinosaur *Coelophysis*. Museum of Northern Arizona Bulletin 57. (2) Whitaker, George O., and Joan Meyers. 1963. Dinosaur Hunt. New York, New York: Harcourt, Brace and World.

32. Petrified Forest National Park

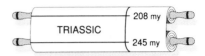

TRIASSIC
208 my
245 my

Holbrook, Arizona

A geologically unstable, gigantic supercontinent straddled the equator in Late Triassic time: Pangea, formed in Permian time as the coalescing land masses of the earth were assembled, was now slowly being sundered by rifting. The modern continents of the world were taking shape. The climate, fauna, and flora throughout Pangea were apparently remarkably uniform, for the fossil plants and animals from sediments of Late Triassic age are similar across the northern hemisphere today.

The dominant land animals in Triassic time were the parareptiles, the lizard- and crocodile-like reptiles. The fossil record reveals, however, that dinosaurs, birds, mammals, and marine reptiles were present by Late Triassic time. Mammals remained minor components in the environments in which they lived; dinosaurs, on the other hand, became increasingly important. At the end of the Triassic, a wave of extinction eliminated the dominant parareptiles and impoverished the ecosystem. The dinosaurs were relatively untouched.

The Chinle Formation is Late Triassic in age. At Petrified Forest National Park, it is made up of variegated, fossil-rich clays (the Petrified Forest Member) that grade upward into red beds. The sediments were deposited by streams that flowed northward and westward across a large, verdant plain located in what is now the Four Corners region of the United States. Discontinuous highlands surrounded the plain, and the area was drained northwestward into the ancient Pacific Ocean in what is now central Nevada (Figure 68).

Petrified Forest National Park is best known for its accumulation of petrified trees of the conifer, *Araucarioxylon arizonicum*, but remains of 200 species of plants have been found. The fauna, too, is complex and includes freshwater invertebrates, fish, amphibians, and reptiles. Thirty-four species of vertebrates are known; among them, as many as six dinosaurs.

Three main paleoecological communities can be identified within the park. A floodplain-swamp community existed along the waterways:

ferns and cycad-like plants growing where they were able to obtain enough light; amphibians 3 meters (10 feet) long (*Metoposaurus*), crocodile-like reptiles (phytosaurs, such as *Rutiodon*), and fish and lungfish living in or near the water. Beyond the waterways on a stable, old floodplain grew a lowland, closed-canopy forest dominated almost entirely by *Araucarioxylon arizonicum*. The trees were gigantic, up to 60 meters (200 feet) tall, and crowned by branches that formed a closed canopy overhead, effectively preventing light from penetrating to the forest floor. The upland community, in contrast, was much more complex. Many kinds of plants flourished; significantly, the primitive seed-bearing plants, the gymnosperms, were diverse. The most noteworthy vertebrates here were the first, small dinosaurs. One species of dinosaur from the upland community is amply preserved at Ghost Ranch *Coelophysis* Quarry [Site 31].

The climate 225 million years ago, as inferred from the fossils of Petrified Forest National Park, was tropical. It is likely that seasonality was pronounced, that the area experienced monsoon-like wet and dry conditions. Lush swamps in the lowlands persisted year round, but the uplands may have experienced profound fluctuations in moisture levels from season to season. Superimposed on the annual cycle was global drought, and with time, the region became increasingly more arid.

Petrified logs occur in high concentrations in Petrified Forest National Park, preserved within layers of silica-rich sediment. The silica precipitated in the pore spaces in the wood and encased the woody tissue itself. Thus are the logs preserved, a rainbow of colors of agate and jasper. Some of the petrified forest consists of stumps up to 3 meters (10 feet) tall preserved in growth position, but many more logs, accumulated as deadfall and driftwood, were concentrated by high flood waters during the monsoons. The majority of the Petrified Forest consists of such logs eroding out of the Chinle Formation strewn on the badlands landscape.

DIRECTIONS: Petrified Forest National Park is located on Interstate 40 between Gallup, New Mexico and Holbrook, Arizona. It is recommended that eastbound visitors take United States Route 180 from Holbrook to the south entrance of the park and leave the park at the north entrance via Interstate 40; westbound visitors exit Interstate 40 at the north entrance and travel south to United States Route 180, then on to Holbrook and Interstate 40 (Figure 70).

PUBLIC USE: Season and hours: Petrified Forest National Park is primarily a day-use park. It is open year round: daily; June to August from 7:00 A.M. to 8:00 P.M., October to April from 8:00 A.M. to 5:00 P.M., May and September from 7:00 A.M. to 6:00 P.M. Well-developed trails and road-side stops lead to concentrations of fossils within the park. **Fees:** $5.00/vehicle or $2.00/person 7-day permit or $25.00 annual Golden Eagle Permit (inquire about other park permits). Free to children 16 years of age and younger. **Food service:** Restaurants operated by concessioners are available at each entrance to the park. **Recreational activities:** Day hiking, backpacking and back-country camping (free permits must be obtained at one

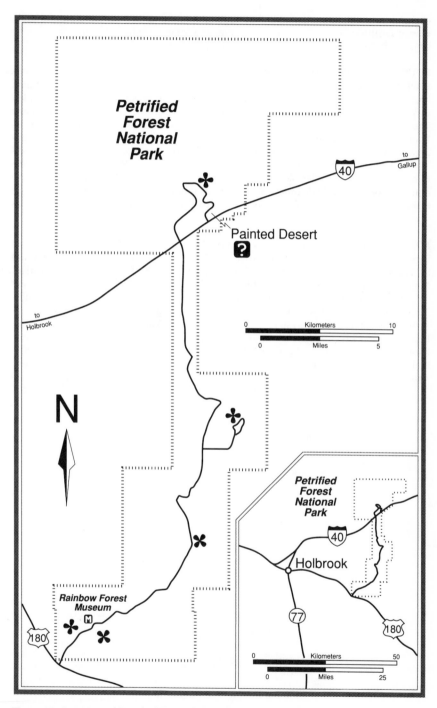

Figure 70. Location of Petrified Forest National Park, Arizona.

of the visitor centers) are available, but no developed camping areas are present. **Handicapped facilities:** The visitor center, museum, and some trails to fossils are accessible by wheelchair. **Restrictions:** Collecting of fossils is prohibited.

EDUCATIONAL FACILITIES: Visitor Center: Painted Desert Visitor Center and Rainbow Forest Museum feature displays and films on various aspects of the geological and human history of the area, the museum featuring the more extensive fossil exhibits. The Painted Desert Inn displays aspects of the culture of southwestern natives. **Visitor Center hours:** The visitor center and museum are open during park hours of operation. **Fees:** None. **Bookstore:** There is a small bookstore at each visitor center featuring books of local and regional interest. **Trails:** Access to fossils is directly from the main north-south park road along numerous short, walking or driving trails that lead into the Petrified Forest. Interpretive information is provided along various portions of the trails and in brochures available at the visitor centers. **Staff programs:** Numerous programs on the paleontology, geology, archeology, and modern ecology of the park are offered on a regular basis, but schedules vary from year to year.

FOR ADDITIONAL INFORMATION: Contact: Superintendent, Petrified Forest National Park, P. O. Box 217, Petrified Forest, Arizona 86028, (602) 524–6228. **Read:** (1) Breed, Carol S., and William J. Breed (editors). 1972. Investigations in the Triassic Chinle Formation. Museum of Northern Arizona Bulletin 47. (2) Colbert, Edwin H., and R. Roy Johnson (editors). 1985. The Petrified Forest through the ages. Museum of Northern Arizona Bulletin 54. (3) Long, Robert A., and Rose Houk. 1988. Dawn of the Dinosaurs: The Triassic in Petrified Forest. Petrified Forest, Arizona: Petrified Forest Museum Association. (4) Jacobs, Louis L., and Phillip A. Murray. 1980. The vertebrate community of the Triassic Chinle Formation near St. Johns, Arizona. Pp. 55–71 in Jacobs, Louis L. (editor). Aspects of Vertebrate History: Essays in Honor of E. H. Colbert. Flagstaff, Arizona: Museum of Northern Arizona Press.

33. Grand Canyon National Park

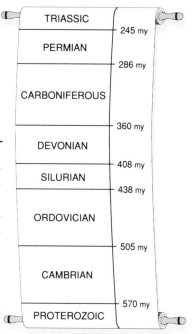

Grand Canyon Village, Arizona

Awesome! The Grand Canyon inspires reverence and wonder, apprehension in the face of nature's grandeur. Even more than the scenery, the geological history exposed here is panoramic, revealed by spectacular erosion that sliced through sedimentary rocks to lay open the bowels of the continent (Figure 71).

The fundamental principles defining geological time are boldly demonstrated in the Grand Canyon. The rocks visible in the canyon walls occur in horizontal layers, one layer deposited on top of another, recording the passage of time—deep time inferred from the layers of rocks. Discrete groups of fossils occur in the distinct layers of sedimentary rocks and are found to succeed each other in an orderly fashion, the fossils themselves a measure of time. Fossils measure time because they preserve the changes in organisms—evolution—that took place over time, changes that are historical in character, constrained by the genes of ancestors and modified by environmental pressures. At this one place, the changes that alter the face of the earth over time are recorded: physical changes of the earth itself, biological changes of the organisms that lived there, and the environmental changes as inferred from the physical and biological.

The view of geological time from the rim of the Grand Canyon is unparalleled. Descent into the canyon is a journey back through time into the earth's distant past. The layered rocks alone, those that make up the canyon walls, span almost 400 million years; but, in the depths of the canyon, igneous and metamorphic rocks—granites and schists—are all that remain of primordial mountains 1.75 billion years old.

The Colorado River flows 1.6 kilometers (1 mile) below the rim of the canyon it has cut. Only at that depth are the rocks of greatest antiquity

217

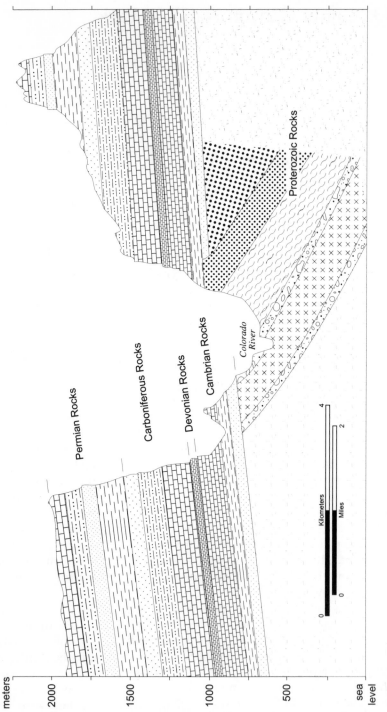

Figure 71. Cross-section through the Grand Canyon.

Permian Rocks

Carboniferous Rocks

Devonian Rocks

Cambrian Rocks

Colorado River

Proterozoic Rocks

meters

2000

1500

1000

500

sea level

Kilometers

Miles

0

2

4

to be found. The granites (Zoroaster Granite) and schists (Vishnu Schist) formed the heart of an ancient mountain chain that stood tall and massive against the edge of the nascent continent of North America. For approximately 600 million years the mountains were scoured and diminished by erosion.

About 1.1 billion years ago deposition began. A thick sequence of sedimentary rocks was deposited in a deep ocean basin on the margins of the continent. The area was subsequently uplifted and tilted by tectonic processes, and many of the rocks were eroded. The layered rocks that remain are called the Grand Canyon Supergroup. Equivalent to the Belt-Purcell Supergroup of Waterton-Glacier International Peace Park [Site 5], the rocks of the Grand Canyon Supergroup contain abundant stromatolites visible in the Bass Limestone.

The western margin of North America achieved a degree of tectonic stability by the end of Precambrian time. Layers of sediments, the horizontal strata of the canyon walls, were deposited; 1600 meters (5200 feet) thick and richly fossiliferous, they accumulated during the Paleozoic Era. The geological history of that time is rich with details.

The first epicontinental sea of Phanerozoic time began to spread over the continent early in the Cambrian; mountains no longer existed on the western margin of proto-North America to impede the flood. As the sea advanced, the waters became deeper, and a sequence of sediments that now make up the Tapeats Sandstone, Bright Angel Shale, and Muav Limestone were deposited (Figure 71). Fossils in the sandstone include trilobites, worm burrows, and animal tracks (good examples are visible at Plateau Point); in the shale are various trace fossils, trilobites, and brachiopods (near Indian Garden, for example); in the limestone, trilobites and brachiopods are most common. Ordovician and Silurian time is not represented in the Grand Canyon, the rocks having been removed by erosion. Similarly, many Devonian rocks have been eroded, but the Temple Butte Limestone is preserved as remnants of a once extensive limestone deposited in an epicontinental sea.

During the Early Carboniferous, another extensive epicontinental sea flooded most of North America, and more limestone was deposited. The Redwall Limestone, like its equivalent the Mission Canyon Limestone at Minnetonka Cave [Site 14], is replete with crinoids, bryozoans, corals, and brachiopods (visible, for example, in the large boulders strewn around at Indian Garden). In the Late Carboniferous and Early Permian, the epicontinental seas gradually withdrew; estuarine and deltaic deposits of the Supai Formation accumulated. The transition is marked by shelly fossils at the base of the formation; petrified wood, vertebrate bones and traces higher up (trails on the Supai redbeds feature displays of vertebrate bones). The upper surface of the Supai Formation is an erosional surface on which the Hermit Shale was deposited.

Evidence of semi-arid conditions at the time of deposition of the

Hermit Shale derives from the fossils: plants such as ferns and conifers, insects, traces of reptiles and amphibians (exhibited at Cedar Ridge). The Coconino Sandstone that overlies the Hermit Shale is a deposit of sand dunes more than 100 meters (330 feet) thick that formed during the most extreme desert conditions known in the history of North America. Vertebrate footprints are the usual fossils from the sandstone (on display at Yavapai Museum).

The uppermost layer of rocks in most places of the Grand Canyon is the Kaibab Limestone. Epicontinental seas returned once again in the Late Permian; the alternating marine limestone and sandstone indicate that the shoreline was very close by. The fossils include new species of clams, snails, corals, brachiopods, crinoids, and sponges. They are readily visible in the canyon walls along the trails.

Triassic deposits are preserved only sparsely in the Grand Canyon area; their wide-spread absence is clear evidence of an unconformity, a long time of erosion that continues in the present. That layers of sediments were laid down in the Mesozoic and Cenozoic is attested to by uneroded rocks elsewhere, but the present episode of erosion in the area of the Grand Canyon has been unrelenting.

The Grand Canyon itself is a young geological feature formed in the last few million years of earth history. The Colorado River cut deeper and deeper into the earth, exposing the horizontal layers of sedimentary rocks. Erosion is the final chapter of the geological history of the area.

Grand Canyon National Park has been designated a UNESCO World Heritage Site in recognition of the grandeur of its scenery and the view it offers into the abyss of time.[19]

DIRECTIONS: Access to Grand Canyon National Park, South Rim is via United States Route 180 (and State Route 64) north from Flagstaff, 130 kilometers (80 miles); via State Route 64 north from Williams, 90 kilometers (56 miles); or via State Route 64 west from Cameron, 85 kilometers (53 miles) to Grand Canyon Village (Figure 72). To the North Rim access is via State Route 67 south from Jacob Lake (and junction with United States Route 89), a distance of 70 kilometers (44 miles).

PUBLIC USE: Season and hours: One can visit Grand Canyon National Park at either the South Rim or the North Rim. The South Rim is open year round, but the range of facilities and activities available decreases during the winter (Labor Day to Memorial Day). The North Rim is open from mid-May to late October. Fossil-bearing rocks and fossils *in situ* are exposed in the walls of the canyon. Weather conditions may limit access to and visibility of outcrops. **Fees:** $10.00/vehicle daily, $4.00/person (for those entering the park by bus, bicycle, or on foot) daily, or $25.00 annual Golden Eagle Permit (inquire about other park permits). **Food service:** Restaurants and stores operated by concessioners are available at both the North and South rims. **Recreational activities:** A full range of park facilities and activities is available including camping (for which an additional fee is charged), hiking, back-country camping (free permit required). Concessioners provide additional activities

---

[19]This famous phrase is from John Playfair, 1802, *Illustrations of the Huttonian Theory of the Earth*. He began a long tradition of popular writing about deep time.

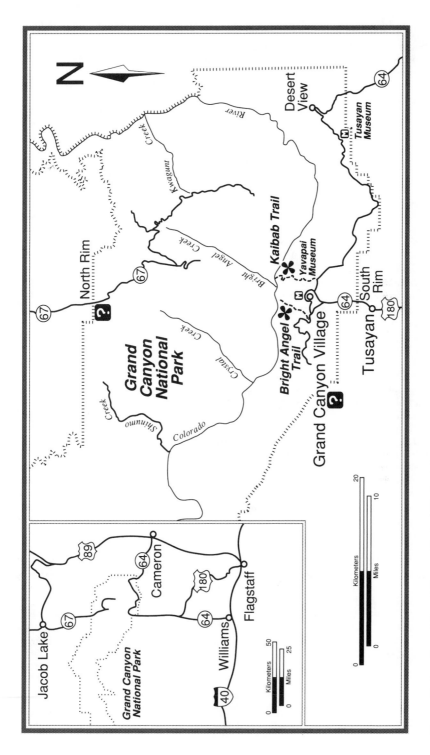

Figure 72. Location of Grand Canyon National Park, Arizona.

such as horseback riding and mule rides into the canyon. **Handicapped facilities:** Many developed areas of the park, including camping facilities, are accessible by wheelchair. **Restrictions:** Collecting of fossils is prohibited. **Transportation:** During the summer season, automobile access to certain points of interest, such as the West Rim Drive to Hermits Rest, is restricted because public use of the park is so high. Instead, the park operates a free shuttle bus service on the South Rim. The West Rim Shuttle operates every 15 minutes from 7:30 A.M. to sunset. There is a free Village Loop Shuttle that operates throughout the Village area from 6:30 A.M. to 9:30 P.M. A free shuttle is also available twice each morning to the Kaibab trailhead.

EDUCATIONAL FACILITIES: Museum: Yavapai Geologic Museum is a small museum dedicated to interpreting the geological history, including paleontology, of the Grand Canyon. Displays focus on the vastness of geological time represented in the Grand Canyon, the use of fossils in interpretation, and on the erosion that has cut the canyon itself. The Tusayan Museum at Desert View exhibits aspects of native American culture and includes a self-guiding trail around a prehistoric dwelling. **Museum hours:** The Yavapai Geologic Museum is open year round: daily; from 9:00 A.M. to 5:00 P.M.; with extended hours in summer from 8:00 A.M. to 8:00 P.M. (times may vary from year to year). The Tusayan Museum open year round: daily; from 9:00 A.M. to 5:00 P.M. **Visitor Center:** Grand Canyon South Rim Visitor Center offers exhibits, dioramas, and films. It is the primary source of visitor information within the park. In addition, the visitor center houses a museum collection of natural and historical objects associated with the Grand Canyon region and a research library. Both facilities are accessible to researchers; the library is also open to museum visitors for in-house use only. **Visitor Center hours:** The visitor center is open year round: daily; from 8:00 A.M. to 5:00 P.M.; with extended hours in summer from 7:30 A.M. to 8:30 P.M. (times may vary from year to year). **Fees:** None. **Bookstore:** Bookstores are present at the museum and at the visitor center. Books, guides, maps, photographic slides, and gift items are available. A book sales area is also present in a restored portion of the Kolb Studio. **Trails:** Fossil-bearing rocks can be seen by hiking along any of the trails, marked and unmarked, that lead into and explore various portions of the Grand Canyon. These are steep and difficult trails requiring forethought and preparation before beginning the descent. The most popular trails are the Bright Angel Trail and the South Kaibab Trail, which include exhibit areas along the trails (for example, Fossil Fern Exhibit along the Kaibab Trail to Cedar Ridge), beginning on the South Rim; the North Kaibab Trail, on the North Rim. In contrast, the Rim Trail is an easy-to-walk trail on the South Rim that extends for 14 kilometers (9 miles) between major points of interest in the park including the museum and visitor center. **Staff programs:** Park naturalists conduct a full range of interpretive nature programs, including a number on geology and fossils (schedules available at all points of information within the park).

FOR ADDITIONAL INFORMATION: Contact: Superintendent, Grand Canyon National Park, P. O. Box 129, Grand Canyon, Arizona 86023, (602) 638-7888. **Read:** (1) Albritton, Claude C., Jr. 1980. The Abyss of Time: Changing Conceptions of the Earth's Antiquity after the Sixteenth Century. San Francisco, California: Freeman, Cooper and Company. (2) Bues, Stanley S., and Michael Morales (editors). 1990. Grand Canyon Geology. New York, New York; Oxford, England: Oxford University Press and Museum of Northern Arizona Press. (3) Thayer, Dave. 1986. A Guide to Grand Canyon Geology along Bright Angel Trail. Grand Canyon Natural History Association.

34. Escalante Petrified Forest State Park

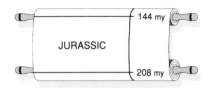

144 my

JURASSIC

208 my

Escalante, Utah

The Late Jurassic Morrison Plain was home to a wide range of plants and animals. The bones of dinosaurs are famous, but plant fossils are rare. Nonetheless, some trees were fossilized in what is now south-central Utah.

The Morrison Plain, some 150 million years ago, was flat, arid, sometimes boggy, and sparsely vegetated except along major water courses. Groves of trees grew on the wet floodplains and in swamps along the few permanent streams that drained the plain. Trees were uprooted during floods and at peak water flow and were carried downstream along with the deadfall; when the water flow waned, the trees were deposited on pointbars and sandbars. Only rapid burial preserved the trees of the Escalante Petrified Forest; and groundwater, rich in silica from volcanic ash, petrified the wood. In contrast, in areas where dinosaur fossils are common (Cleveland-Lloyd Dinosaur Quarry [Site 39], Dinosaur National Monument [Site 40]), plant remains were rapidly decomposed.

The plants of the Morrison Formation are poorly preserved and poorly understood, and only broad ecological interpretations are possible. It appears that conifers were distributed along the waterways, ferns grew in moist areas where there were few trees and light was adequate, and scrubby cactus-like cycads capped the higher ground. When the rains came, luxuriant growth quickly replaced the desert scrub trees. Yet vegetation was ephemeral, for the climate was monsoonal, controlling the patterns of life and, often, death.

At Escalante Petrified Forest State Park, petrified wood is weathered out of the softer sediments which entomb it. In some places, slivers of wood litter the ground. In others, chips of petrified wood bear evidence of having been worked by human hands. Still elsewhere in the park, large segments of tree trunks are preserved.

DIRECTIONS: Escalante Petrified Forest State Park is located near the town of Escalante. Follow State Route 12 for 1.6 kilometers (1 mile) west of the town of Escalante, then proceed north on a gravel road 0.8 kilometer (0.5 mile) to the park gate (Figure 73). The parking lot

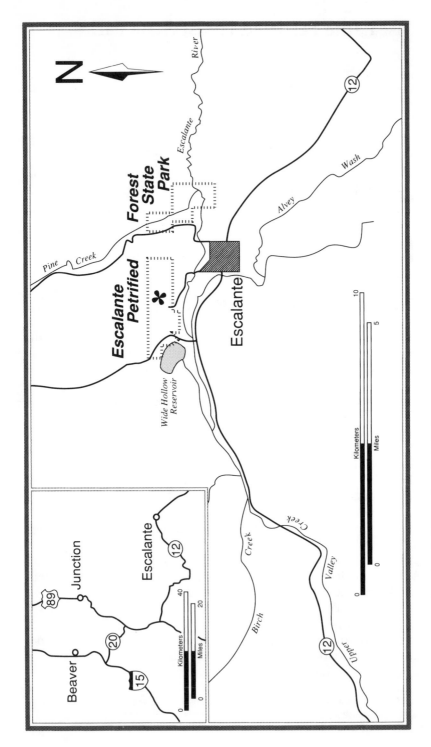

Figure 73. Location of Escalante Petrified Forest State Park, Utah.

and trail head are 0.4 kilometer (0.25 mile) beyond the gate. Access to a second site is due north of Escalante.

PUBLIC USE: Season and hours: The Escalante Petrified Forest State Park is open year round as are the trails that provide access to fossils. **Fees:** $3.00/vehicle daily or annual Utah State Park Permit. **Food service:** Restaurants and stores are available in Escalante. **Recreational activities:** Boating, fishing, swimming, and camping (for which a fee is charged) are available. **Restrictions:** Collecting of fossils is prohibited.

EDUCATIONAL FACILITIES: Visitor Center: A new park office and visitor station is being built which will feature displays and local information. **Trails:** The areas of fossil exposure are reached by way of hiking trails that begin at the parking lot. They are easy to walk, but some short portions of the trail are quite steep. One trail is 1.2 kilometers (0.75 mile) long and is self-guiding; the second is a 1.7 kilometer (1 mile) trail that branches from and then rejoins the first.

FOR ADDITIONAL INFORMATION: Contact: Park Ranger, Escalante Petrified Forest State Park, P. O. Box 350, Escalante, Utah 84726, (801) 826–4466.

35. Moab Dinosaur Tracks

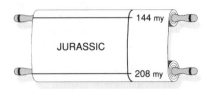

144 my

JURASSIC

208 my

Moab, Utah

Dinosaur tracks can be found in every Jurassic rock formation exposed in eastern Utah. Tracks near Moab are in the Kayenta Formation and are Early Jurassic in age. The rocks, predominantly sandstones, siltstones and claystones, now a rich red in color, have long been known to contain bones of early vertebrates, but only recently have the extensive dinosaur tracks attracted much attention. The rocks of the Kayenta Formation are part of a thick sequence exposed in Colorado, Utah, and New Mexico that were deposited in desert and semi-arid environments as large areas of the western continent became subaerially exposed. The Kayenta Formation itself is predominantly fluvial and playa lake in origin, the sediments associated with ephemeral water.

The tracksite in the Kayenta Formation near Moab preserves the footprints of two different carnivorous dinosaurs (theropods). Both tracks are three-toed, clawed footprints: one large, with a maximum length of 37 centimeters (15 inches) and maximum width of 30 centimeters (12 inches); one small, perhaps made by a coelurosaur, with a maximum length of 14 centimeters (5.5 inches) and a maximum width of 11 centimeters (4 inches). Among them are the tracks of a tiny animal, probably a juvenile of one of the adult forms.

Analysis of the Moab dinosaur tracks and the tracks from nearby sites indicates that there were many track-making animals despite the apparently harsh conditions of a desert. Coelurosaurs, it seems, preferred desert environments. It is also interesting and revealing of their ecological preferences that tracks of the sauropods and ornithopods, herbivores known elsewhere from deposits of similar age, are entirely absent.

The Moab Dinosaur Tracks are visible from a road-side stop part way down the cliff face of the canyon through which the Colorado River flows. A viewing tube mounted by the side of the road orients one in the direction of the tracks. A closer view can be obtained by using binoculars or by scrambling up the steep cliff face.

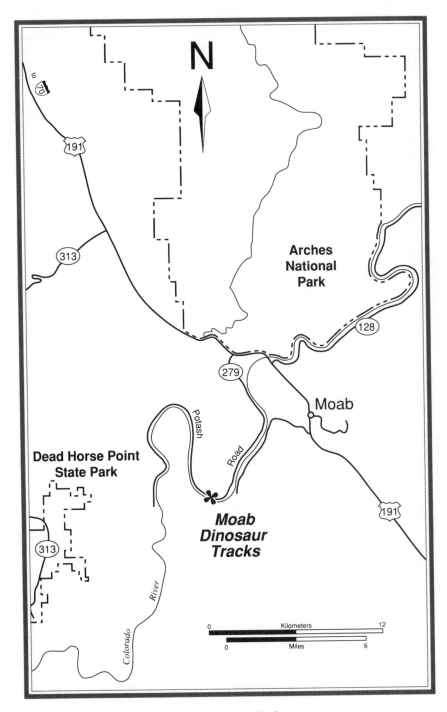

Figure 74. Location of the Moab Dinosaur Tracks, Utah.

DIRECTIONS: The Moab Dinosaur Tracks site is on the west side of the Colorado River along State Scenic Route 279. Access to State Scenic Route 279 (Potash Road) via United States Route 191 is 5.2 kilometers (3.2 miles) north of the Moab Visitor Center; the tracksite is 9.8 kilometers (6 miles) along State Scenic Route 279 (Figure 74). The site is referred to in the Moab Area Auto Tours pamphlet available at the Moab Visitor Center.

PUBLIC USE: Season and hours: The Moab Dinosaur Tracks site is on public land administered by the Bureau of Land Management and is accessible year round. **Fees** None. **Food service:** Restaurants and stores are available at Moab. **Recreational activities:** None on site. **Restrictions**: Collecting of fossils is prohibited.

EDUCATIONAL FACILITIES: Visitor Center: Although it is not a visitor center, the Bureau of Land Management, Moab District Office (located at 82 E. Dogwood) in Moab has a small fossil display and welcomes visitors. It is open during normal business hours.

FOR ADDITIONAL INFORMATION: Contact: Bureau of Land management, Grand Resource Area Office, P. O. Box M (Sand Flat Road), Moab, Utah 84532, (801) 259–8193. **Read:** (1) Lockley, Martin G. 1986. Dinosaur Tracksites: A Guide to Dinosaur Tracksites of the Colorado Plateau and American Southwest. The First International Symposium on Dinosaur Tracks and Traces, Albuquerque, 1986. Geology Department Magazine, Special Issue Number 1, A University of Colorado at Denver, Geology Department Publication. (2) Lockley, Martin G. 1986. The paleobiological and paleoenvironmental importance of dinosaur footprints. Palaios, volume 1, pp. 37–47.

36. Mill Canyon Dinosaur Trail

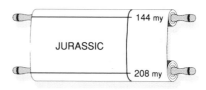

144 my

JURASSIC

208 my

Moab, Utah

The Mill Canyon Dinosaur Trail features an exposure of the Morrison Formation, the highly fossiliferous layer of rocks that is widely exposed in Utah, Colorado, Wyoming, and adjacent states. The rocks were deposited on an ancient lake plain by rivers, isolated lakes, and swamps. Dinosaur bones are common remains in the Morrison Formation and attest to the abundance and diversity of animals that lived 150 million years ago.

At Mill Canyon, the dinosaur bones are scattered and fragmentary; the condition of the bones very unlike that of the bones at other Morrison sites, such as Cleveland-Lloyd Dinosaur Quarry [Site 39] and Dinosaur National Monument [Site 40], where bones are exceedingly well preserved in large numbers. The differences between the fossil localities reflect the ecological variety that was present when the dinosaurs roamed on the Morrison plain and underscore the fortuitous nature of the preservation of all fossils. In the case of Mill Canyon, an animal died on the plain, its carcass was attacked by scavengers, and its bones were disarticulated and scattered; only a few bones chanced to be buried and preserved. In the other cases, natural traps or mass death followed by rapid burial account for spectacular preservation.

A number of dinosaur bones and some fossilized wood are exposed at Mill Canyon along a short self-guiding trail. These fossils have been left *in situ* unaltered, except that some have been prepared and preserved by a paleontologist to insure that they do not deteriorate. The physical appearance of the site and the bones it contains are modest, yet the site accurately represents the conditions under which most paleontological excavation is conducted.

DIRECTIONS: Access to the Mill Canyon Dinosaur Trail via United States Route 191 is 24 kilometers (14.8 miles) north of the Moab Visitor Center in Moab, Utah. Note that railroad tracks extend parallel to United States Route 191 along the west side of the highway in this area. The access road from United States Route 191 is difficult to find. It can be spotted by its location relative to a railroad bridge: it is a dirt road that runs to the west of the highway,

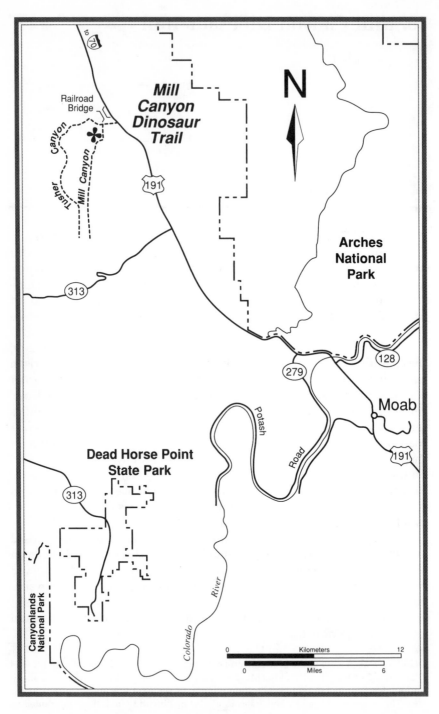

Figure 75. Location of Mill Canyon Dinosaur Trail, Utah.

crossing the railway tracks just south of the railroad bridge (Figure 75). The first posted signs are west of the tracks; from that point the road is well signed. Distance from United States Route 191 to the parking area and visitor box is 3 kilometers (1.8 miles).

PUBLIC USE: Season and hours: Mill Canyon Dinosaur Trail is on public land administered by the Bureau of Land Management. It is open to the public year round. The dinosaur trail is part of the Monitor and Merrimac Mountain Bike Trail (designed for all-terrain vehicles). **Fees** None. **Food service:** Restaurants and stores are available in Moab. **Recreational activities:** None on site. **Restrictions**: Collecting of fossils is prohibited.

EDUCATIONAL FACILITIES: Visitor Center: Although it is not a visitor center, the Bureau of Land Management, Moab District Office in Moab has a small fossil display and welcomes visitors. It is open during normal business hours. **Trails:** The self-guiding trail has 15 closely spaced stops with fossils visible at each stop. It is short, less than 1 kilometer (0.6 mile) one way, and easy to walk. Although the trail is not steep, it is narrow where it has been cut into a steep portion of the hillside.

FOR ADDITIONAL INFORMATION: Contact: Bureau of Land Management, Grand Resource Area Office, P. O. Box M, Sand Flat Road, Moab, Utah 84532, (801) 259–8193.

37. Dinosaur Hill/Riggs Hill Trails

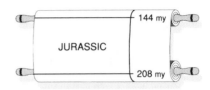

144 my

JURASSIC

208 my

Grand Junction, Colorado

The Dinosaur Hill and Riggs Hill trails near Grand Junction are largely of historical interest. Each trail leads to a site of major discoveries of dinosaurs made at the turn of the century in the Morrison Formation, rocks that are 150 to 140 million years old.

Rocks of the Morrison Formation are widely distributed across Colorado, Utah, Wyoming, and adjacent states. They comprise complex alluvial fan deposits. For approximately 10 million years, sediments were carried eastward and northward by streams, flowing from young mountains in the west and southwest, and deposited in the shallow basin of a drying inland sea, the Sundance Sea. Dinosaurs were plentiful on the Late Jurassic plain for their bones are common in the Morrison Formation today (for example, Cleveland-Lloyd Dinosaur Quarry [Site 39] and Dinosaur National Monument [Site 40]). The early history of vertebrate paleontology in North America is largely a chronicle of discoveries of dinosaur bones in the Morrison Formation.

In 1900, Elmer Riggs of the Field Museum of Natural History in Chicago discovered portions of a dinosaur at the site that now bears his name. That dinosaur was given the name *Brachiosaurus altithorax* and dubbed the largest land animal that ever lived. Riggs returned to Grand Junction in 1901 and, at a nearby site, excavated a skeleton of a giant *Apatosaurus*. That site is now called Dinosaur Hill. Since those initial discoveries, other dinosaurs have been found at Riggs Hill and Dinosaur Hill, but what remains today of the giant skeletons is a portion of the tail, the last 4 or 5 meters (12 or 15 feet), of the *Apatosaurus* and a boulder at Dinosaur Hill bearing a mold of a *Diplodocus* thigh bone. Other bones, however, are regularly discovered in the surrounding area.

DIRECTIONS: Access to each site is via State Route 340: Dinosaur Hill is 2.5 kilometers (1.5 miles) south of Fruita; Riggs Hill is reached by continuing southeast on State Route 340 to South Broadway and then following South Broadway to its intersection with Meadows Way where the trail is located (Figure 76).

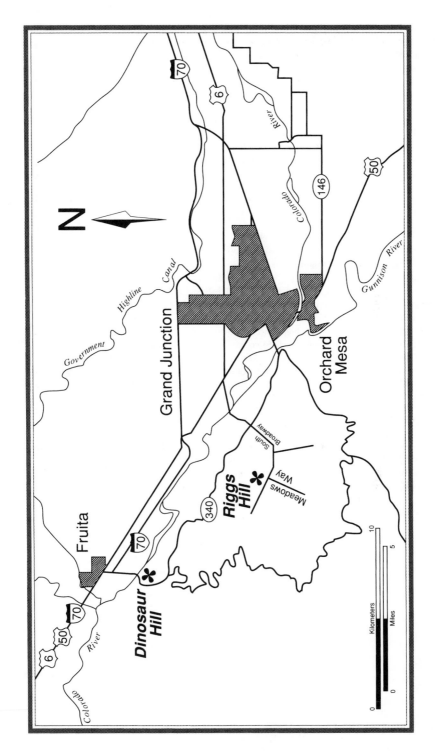

Figure 76. Location of Dinosaur Hill/Riggs Hill Trails, Colorado.

PUBLIC USE: Season and hours: The Riggs Hill and Dinosaur Hill sites are owned by the Museum of Western Colorado. The trails are the result of cooperative efforts by the Bureau of Land Management, Museum of Western Colorado, and community involvement. They are open to the public year round. **Fees** None. **Food service:** Restaurants and stores are available in nearby towns. **Recreational activities:** None on site. **Restrictions:** Collecting of fossils is prohibited.

EDUCATIONAL FACILITIES: Museum: Dinosaur Valley, Museum of Western Colorado, is located in the city of Grand Junction. Dinosaur Hill and Riggs Hill trails are an extension of an innovative concept in museum development. Dinosaur Valley is housed in an old downtown department store building (at 4th and Main). It presents traditional museum displays beside scaled-down mechanical models of dinosaurs. One corner of the building is set aside as an accessible preparation area where visitors can watch preparators at work and ask questions of them. The museum is accessible by wheelchair. **Museum hours:** The museum is open year round: daily in summer, Memorial Day to September 30, from 9:00 A.M. to 6:00 P.M.; Tuesday to Sunday in winter from 10:00 A.M. to 4:30 P.M.. **Fees:** Adults $3.50; children $2.00. **Bookstore:** An extensive giftshop offers a small selection of books of local interest and assorted dinosaur memorabilia. **Trails:** Dinosaur Hill and Riggs Hill trails are short (each is 1.25 kilometers or 0.75 mile), self-guiding trails that lead to historical excavation sites. The fossils one can see are limited to a few dinosaur vertebrae, a mold of a dinosaur bone, and a fossil soil horizon at Dinosaur Hill. The historical quarries have been partially reconstructed and feature cement replicas of dinosaur bones. **Special group activities:** The Museum of Western Colorado has two-day weekend field and laboratory programs which are run every weekend in the summer. The Bureau of Land Management and the Museum of Western Colorado cooperate to open quarry sites form time to time for public participation in excavation. **Note:** Dinamation International Society conducts expeditions or field trips oriented toward collecting dinosaurs. Costs vary with the specifications of each expedition. Contact Dinosaur Discovery Expeditions, 27362 Calle Arroyo, San Juan Capistrano, California 92675, 1–(800) 547–0503.

FOR ADDITIONAL INFORMATION: Contact: Dinosaur Valley, Museum of Western Colorado, Fourth and Main Streets, P. O. Box 20000–5020, Grand Junction, Colorado 81502–5020, (303) 241–9210. **Read:** (1) Averett, Walter R. (editor). 1987. Paleontology and Geology of the Dinosaur Triangle: Guidebook for the 1987 Field Trip, September 18–20, 1987. Grand Junction, Colorado: Museum of Western Colorado. (2) Callison, George, and Helen M. Quimby. 1984. Tiny dinosaurs: are they fully grown? Journal of Vertebrate Paleontology, volume 3, number 4, pp. 200–209. (3) Colbert, Edwin H. 1968. Men and Dinosaurs. New York, New York: Dutton.

38. Rabbit Valley Trail Through Time

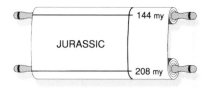

144 my

JURASSIC

208 my

Mesa County, Colorado

It seems that wherever the Morrison Formation is exposed dinosaur bones are to be found. So it is at Rabbit Valley in western Colorado where the bones of dinosaurs occur as isolated deposits of partial skeletons. The bones, carried downstream by one of the many Late Jurassic rivers as it flowed through the region, became widely dispersed. At Rabbit Valley, 10 genera have been recovered, including *Stegosaurus*, *Camarasaurus*, and *Allosaurus*. In addition to dinosaur bones are the bones of crocodiles and turtles.

The Rabbit Valley Trail Through Time illustrates the fortuitous aspect of fossil preservation in the Morrison Formation. The fossils are scattered and isolated, apparently randomly distributed throughout the sediments. In this feature, the preservation is similar to that at Mill Canyon Dinosaur Trail [Site 36] and is in sharp contrast to the spectacular accumulation of bones at Dinosaur National Monument [Site 40]. Why is there such a difference?

It appears that some dinosaurs were solitary and lived on the drier, upland areas of the Morrison plain; *Stegosaurus*, for example. When a stegosaur died, its carcass was attacked by scavengers and its bones scattered. The chances were small that a few bones would became buried and be preserved. Such solitary animals that lived in upland environments on the Morrison plain are much less well represented than are herd animals that lived along river valleys (for example, the sauropod herds preserved at Dinosaur National Monument [Site 40]).

The Trail through Time is located in the Rabbit Valley Research Natural Area. Bones of dinosaurs and plant remains are exposed *in situ* in the rock.

DIRECTIONS: The Rabbit Valley Natural Research Area located on Interstate 70 is 3.5 kilometers (2 miles) east of the Utah border and 50 kilometers (30 miles) west of Grand Junction, Colorado (Figure 77). Take Exit 2 and cross to the north side of Interstate 70, where a parking area is available. The trail begins at the parking area.

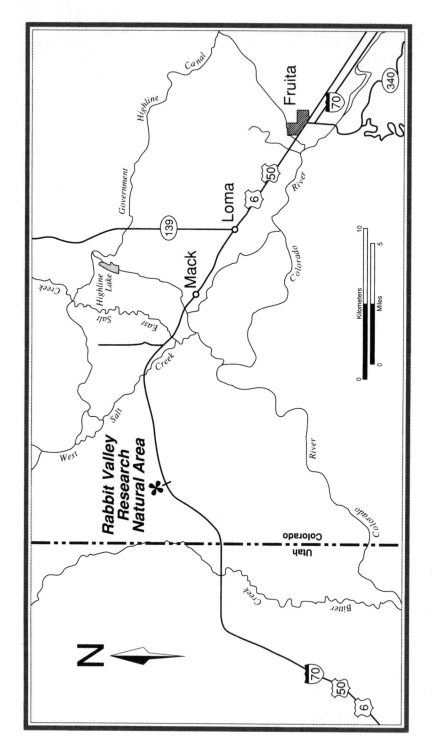

Figure 77. Location of Rabbit Valley Trail Through Time, Colorado.

PUBLIC USE: Season and hours: The Rabbit Valley Research Natural Area is located on public lands managed by the Bureau of Land Management. It is accessible to public visitation year round. The area was developed with the cooperation of the Museum of Western Colorado in Grand Junction. **Fees:** None. **Food service:** Restaurants and stores are available in the Grand Junction area. **Recreational activities:** None on site. **Restrictions**: Collecting of fossils is prohibited.

EDUCATIONAL FACILITIES: Trails: The Trail Through Time is short (about 1.6 kilometers or 1 mile round trip), self-guiding (trail guides provided), well-marked, and easy-to-walk. Geological features are emphasized to provide background information about the environment of the area at the time the bones were deposited. **Special group activities:** Museum of Western Colorado in Grand Junction and the Bureau of Land Management cooperatively offer a variety of interpretive, public education, and public participation programs including excursions to the Rabbit Valley area. Contact the Museum of Western Colorado for detailed information.

FOR ADDITIONAL INFORMATION: Contact: Bureau of Land Management, Grand Junction District Office, 764 Horizon Drive, Grand Junction, Colorado 81506, (303) 243–6552 *or* Museum of Western Colorado, P. O. Box 20000–5020, Grand Junction, Colorado 81502–5020, (303) 242–0971. **Read:** (1) Averett, Walter R. (editor). 1987. Paleontology and Geology of the Dinosaur Triangle: Guidebook for the 1987 Field Trip, September 18–20, 1987. Grand Junction, Colorado: Museum of Western Colorado. (2) Keener, James. 1989. Dinosaur Triangle: Land of the Terrible Lizards. Grand Junction, Colorado.

39. Cleveland-Lloyd Dinosaur Quarry

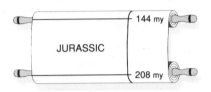

144 my

JURASSIC

208 my

Price, Utah

The rocks of the Morrison Formation are known around the world for the prodigious numbers of dinosaur bones they bear. Among the richest and most unusual of the fossil localities is the Cleveland-Lloyd Dinosaur Quarry in central Utah. The accumulation of bones is particularly dense, unusual because most of the remains are of the carnivore, *Allosaurus*.

Sea level in Late Jurassic time was considerably higher on a global scale than it is today, and large areas of continents were flooded by extensive inland seas. In western North America, the Sundance Sea extended across what is now Montana, Wyoming, Colorado, Utah, and adjacent states. It was separated from open ocean to the west by a chain of mountains in Nevada and Idaho; the Nevadan Mountains had been built earlier in Jurassic time. As the mountains continued to rise and the sediments from them were carried eastward to be deposited, the Sundance Sea retreated northward leaving behind an expanse of arid lowland plain in the rainshadow of the mountains. Rivers, lakes, and swamps punctuated the flat landscape, the water supply controlled by seasonal and monsoonal rains. The sands, silts, and muds that were deposited are now the variegated rocks of the Morrison Formation (Figure 78).

Dinosaurs were the megafauna of the Morrison plain. Large herbivorous dinosaurs predominated. As in modern ecosystems, carnivores, in particular the large carnivores, were much less numerous than their prey; in most fossil deposits, bones of large carnivores are similarly much less numerous. But not so at the Cleveland-Lloyd Dinosaur Quarry, where at least 44 skeletons of *Allosaurus* have been recovered.

The site of the Cleveland-Lloyd Dinosaur Quarry was a dinosaur death trap. Carnivores were attracted to the site over a period of time by unusual circumstances; they became trapped and died, and their accumulated bones were preserved. It appears that about 150 million years ago, a small pond on the floodplain of a river was a watering hole on the Morrison plain during the dry season. The pond probably had a quicksand base because the large herbivores it attracted became bogged

238

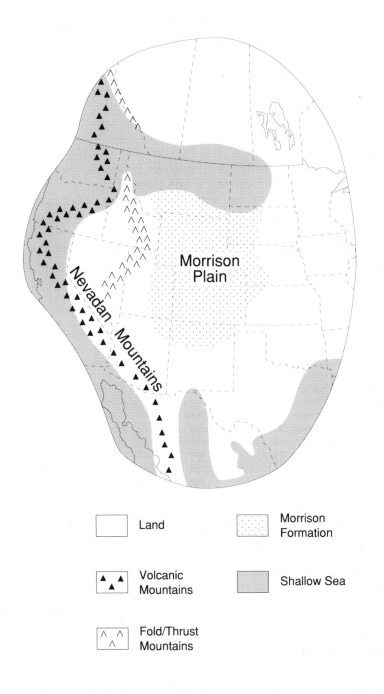

Figure 78. Paleogeography of western North America during Late Jurassic time. The Sundance Sea had dried and retreated northward, and the sediments now called the Morrison Formation were deposited on the Morrison plain.

down and were unable to escape. Their carcasses, an abundance of food, attracted carnivores, which in turn became trapped. The predator became the prey and attracted more carnivores to the pond; the bones of the dead animals became jumbled and broken by the animals scavenging on the surface. When the pond dried the animals stopped coming, and sediments covered the trap.

Accumulations of carnivore bones are rare in the fossil record for they require that a mechanism existed, while the animals lived, to concentrate carnivores in a small area. Cleveland-Lloyd Dinosaur Quarry, however, is not unique in that respect for it is comparable with the Rancho La Brea tarpits at Hancock Park [Site 43].

Of the dinosaurs recovered at the Cleveland-Lloyd Dinosaur Quarry, the majority are *Allosaurus*, large bipedal carnivores—the state fossil of Utah. Among the others are 13 herbivores, including the rare *Stegosaurus*, and six other carnivores. This rich site has been recognized as a United States National Natural Landmark. Specimens from the quarry are on display in museums around the world.

DIRECTIONS: Access to the Cleveland-Lloyd Dinosaur Quarry is circuitous, but there are many signs along the route. The quarry is located approximately 45 kilometers (30 miles) south of Price via State Route 10 and State Route 155, through the town of Elmo or the town of Cleveland and the Desert Lake Waterfowl Management Area, and eastward along a graded road that traverses open range land (Figure 79). The following description details the most direct route from Price. Proceed south from Price on State Route 10 for 17.5 kilometers (10.8 miles) to the junction of State Route 155. Follow State Route 155 for 4 kilometers (2.5 miles); an unnamed county road leads east-southeast for a distance of 3.2 kilometers (2 miles) to the town of Elmo (State Route 155 continues south to the town of Cleveland). As one enters Elmo, there is a stop sign at West Elmo Road; proceed south one town block to Main Street and follow Main Street east for 1.6 kilometers (1 mile) through Elmo and out of town to a graded, gravel road that leads to the south. There are signs at various distances along this road directing one to the quarry site. Note that the road proceeds in an easterly direction as it crosses the Desert Lake Waterfowl Management Area. At a distance of 5.8 kilometers (3.6 miles) along the graded road, a quarry sign indicates a right turn (the left fork leads to Wellington); 2.5 kilometers (1.5 miles) farther another quarry sign indicates a left turn; 8.5 kilometers (5.3 miles) farther there is a fork in the road and a quarry sign indicates a left turn to the quarry gate (and registration box) which is 1.4 kilometers (0.8 mile) beyond.

PUBLIC USE: Season and hours: The Cleveland-Lloyd Dinosaur Quarry is managed and operated by the Bureau of Land Management. It is open Easter to Labor Day: weekends in spring, Easter to Memorial Day, from 10:00 A.M. to 5:00 P.M.; Thursday to Monday in summer, Memorial Day to Labor Day, from 10:00 A.M. to 5:00 P.M. **Fees:** None. **Food service:** Restaurants and stores are available in towns. **Recreational activities:** Picnicking is available on site. **Handicapped facilities:** The visitor center and quarry building are accessible by wheelchair. **Restrictions**: Collecting of fossils is prohibited.

EDUCATIONAL FACILITIES: Visitor Center: A visitor center on site presents geological and paleontological information and features displays of *Allosaurus*, in particular, a mounted skeleton that towers above the visitor. The quarry site is enclosed in a separate building which is open to visitors. Even if excavation is not in progress at the time of a visit, *in situ* bones are visible and various excavating techniques can be seen. An interpre-

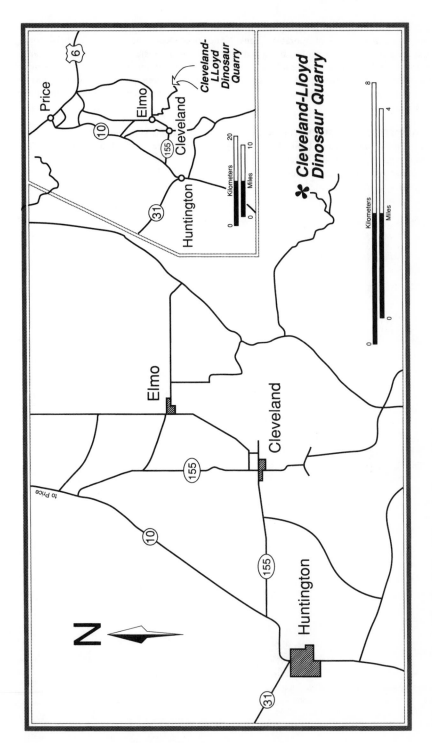

Figure 79. Location of Cleveland-Lloyd Dinosaur Quarry, Utah.

tive guide is present while the site is open. **Visitor Center hours:** The visitor center is open during the quarry hours of operation. **Fees:** None. **Bookstore:** There is no true bookstore or gift shop at the visitor center, but a number of items of special interest to the area are available for sale. **Trails:** A self-guiding trail leads from the visitor center across exposures of the Morrison Formation, past bones preserved *in situ* outdoors, past the enclosed quarry site, and returns to the visitor center. It is approximately 3 kilometers (2 miles) long and easy-to-walk.

FOR ADDITIONAL INFORMATION: Contact: Bureau of Land Management, Price District Office, 700N 700E, Price, Utah 84501, (801) 637–4584. **Read:** (1) Madsen, J. H. 1976. *Allosaurus fragilis:* A Revised Osteology. Utah Geological and Mineralogical Survey, Bulletin 109. (2) Keener, James. 1989. Dinosaur Triangle: Land of the Terrible Lizards. Grand River, Colorado.

40. Dinosaur National Monument

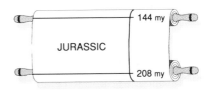

144 my

JURASSIC

208 my

Jensen, Utah

Dinosaur National Monument is the ultimate *in situ* dinosaur exhibit. Hundreds of bones of enormous dinosaurs, carefully prepared, are exposed on a sloping wall of rock for all visitors to see and marvel. Preparators work to reveal the bones, removing only those that lie on the top of the main bone-bearing layer.

The dinosaur bonebed is part of the 150 to 140 million year old Morrison Formation, multi-colored rocks made up of sands, silts, and clays that were deposited on a lake plain in Late Jurassic time (Figure 78). In earlier Jurassic time, two geological features had characterized the southwestern quadrant of North America: one, the Nevadan mountains being elevated in Nevada and Idaho; the other, the Sundance Sea, an epicontinental sea extending across present day Montana, Wyoming, Colorado, and Utah and large portions of adjacent states. Sediment was carried eastward by rivers from the mountains and deposited in the sea; the sea retreated northward in response.

The Morrison plain lay in the rainshadow of the mountains; its climate, seasonal and monsoonal. Rivers flowed sluggishly in the dry season, torrentially in the wet; ponds and lakes evaporated in the dry season, becoming stagnant and saline, only to be recharged again when the rains came. Large numbers of dinosaurs roamed the plain; among them were sauropods, the largest of the dinosaurs.

The fantastic collection of dinosaur bones at Dinosaur National Monument was accumulated by seasonal floods. During a flood, torrential rivers picked up the dead and decayed carcasses that were scattered on the floodplain, grim testimony to drought, and carried them downstream. Many of the bones were deposited high on a sandy point bar in the river and, as the flood waters receded, were covered with a layer of silt and mud. Three times flood waters brought a bedload of dinosaur bones to rest on the point bar; each time a representative sample of the dinosaur fauna was preserved.

The accumulation of fossils at Dinosaur National Monument represents the natural mortality of extant populations and, therefore, documents the number of species that were living in the area at the time and the relative numbers of each. The major groups of Late Jurassic dinosaurs are all represented, at least 100 individuals of 10 different species, but not all equally well. The most common are the sauropods; the most common sauropod, *Camarasaurus*. They apparently lived and died on the floodplain and near the river. Dinosaurs such as ornithopods, the bipedal herbivores, that lived farther away from the river are less well represented; their bones are isolated, fewer in number, and many show the ravages of prolonged exposure to weathering. Carnivorous dinosaurs, such as the Cleveland-Lloyd *Allosaurus* [Site 39], are rare. Some dinosaurs lived too far away from the river for their bones to be gathered by the flood waters; they are unknown at Dinosaur National Monument.

The giants exposed on the wall command the attention of visitors to the monument. The Morrison Formation, however, is rich in fossils throughout its extent and is yielding bones of the minuscule—frogs, salamanders, and clams; even mammals!—as well as the gigantic. They add finer details to the picture of life on the Morrison plain.

Dinosaur National Monument is known for its man (and now woman) on the wall, preparators at work exposing dinosaur bones in the near vertical wall of rock around which a visitor center has been built. The bones are exposed, but most are not removed, so that visitors can see them as they were when deposited on an ancient sandbar (Figure 80). The number of bones exposed on the 50 meters (150 feet) wide by 15 meters (50 feet) high sandstone wall is estimated to be between 2000 and 3000.

DIRECTIONS: Dinosaur National Monument straddles the Colorado-Utah boundary and can be reached via United States Route 40 (Figure 81). The Dinosaur Quarry itself is in the Utah portion of the park. Follow State Route 149 north from United States Route 40 at Jensen for 11 kilometers (7 miles) to the main parking lot and the shuttle service. In winter proceed another 1 kilometer (0.6 mile) to the quarry building.

PUBLIC USE: Season and hours: Access to fossils at Dinosaur National Monument is at the quarry, which is permanently enclosed by the visitor center. The quarry is open year round: daily; from 8:00 A.M. to 7:00 P.M.; except closed New Year's Day (January 1), Thanksgiving Day, and Christmas Day (December 25). The park is open year round. **Fees:** None to the park. To the quarry: $5.00/vehicle or $2.00/person daily or $25.00 annual Golden Eagle Permit (inquire about other park permits). **Food service:** Restaurants and stores are available in Dinosaur, Jensen, and Vernal. **Recreational activities:** A variety of activities are available in the park including hiking, river running, fishing, and camping (for which a fee is charged at some sites). **Handicapped facilities:** The visitor center and park headquarters are accessible by wheelchair. **Restrictions:** Collecting of fossils is prohibited. **Transportation:** During periods of peak visitation visitors are required to leave their vehicles at the main, lower parking lot and take the shuttle bus to the quarry building. The shuttle is free and is scheduled to meet demand.

EDUCATIONAL FACILITIES: Visitor Center: The visitor center was built to enclose the steeply inclined layer of sandstone within which the dinosaur bones are preserved. The primary attraction is the wall of fossils, but interpretive displays, observation of the paleon-

Figure 80. The wall of fossils at Dinosaur National Monument. The upper photograph illustrates a section of the wall with its *in situ* fossils enclosed by the visitor center; the lower photograph, a detail of some of the dinosaur bones.

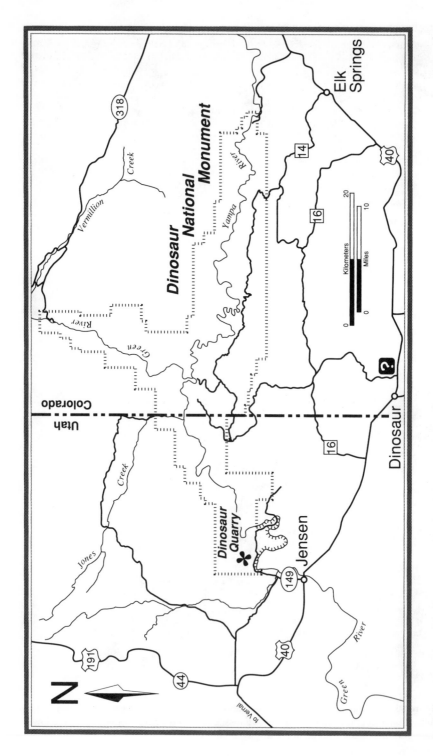

Figure 81. Location of the Dinosaur Quarry in Dinosaur National Monument, Colorado and Utah.

246

tology laboratory, and talks given by park staff complement the *in situ* exhibit. Park headquarters at Dinosaur has a visitor center with exhibits and an audiovisual program. **Visitor Center hours:** The visitor center is open during quarry hours of operation. **Fees:** The quarry entrance fee cited above is inclusive. **Bookstore:** A small bookstore and gift shop is present at the quarry visitor center. It concentrates on items that relate to the fossils and natural history of the area. **Tour guide:** Tour guides are available to special interest groups on request. **Interpretive sign:** There are outdoor interpretive panels at the main parking lot at the base of the quarry and at the park headquarters. **Staff programs:** Interpretive programs about fossils are included as part of the regular summer programming. Schedules are posted.

FOR ADDITIONAL INFORMATION: Contact: Superintendent, Dinosaur National Monument, Headquarters, P. O. Box 210, Dinosaur, Colorado 81610, (801) 789–2115. **Read:** (1) Averett, Walter R. (editor). 1987. Paleontology and Geology of the Dinosaur Triangle: Guidebook for the 1987 Field Trip, September 18–20, 1987. Grand Junction, Colorado: Museum of Western Colorado. (2) Keener, James. 1989. Dinosaur Triangle: Land of the Terrible Lizards. Grand River, Colorado.

41. Berlin-Ichthyosaur State Park

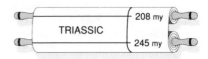

TRIASSIC — 208 my / 245 my

Austin, Nevada

Pangea was only just beginning to break up in Late Triassic time. The portion of that landmass that is now North America was situated in the tropics north of the equator, and its Pacific margin formed a portion of the long western coastline of Pangea that extended virtually from pole to pole. The Cordilleran Mountains that frame the west coast of North America today did not exist, although the highlands of the Antler Orogeny still stood tall. Volcanic island archipelagos lay offshore; rich, shallow seas separated the islands from the mainland, spilling over onto the continent from time to time. Marine reptiles were ubiquitous in the coastal waters adjacent to the Pangea; their distribution in North America extended, in present terms, from California and Nevada to the Canadian Arctic Islands.

The Late Triassic was a time of dramatic evolutionary experimentation among reptiles. On land, for example, dinosaurs emerged (Ghost Ranch *Coelophysis* Quarry [Site 31], Petrified Forest National Park [Site 32]). But equally important innovations took place in the oceans where many groups of reptiles adopted distinctive anatomical modifications that allowed them to exploit life in the water in unique ways.

Among the most specialized marine reptiles were the ichthyosaurs, fish-lizards with a body shape as streamlined as that of modern tuna. The streamlined body, with elongate head and shortened neck, and the strong tail testify to the power with which ichthyosaurs could swim. Limbs were modified to form flippers. The huge eyes imply keen eyesight; long jaws with sharp, conical teeth indicate carnivory. These animals, the porpoises of the Mesozoic world, were powerful swimmers and active carnivores. Their total transformation to a marine existence is confirmed by fossils of adult individuals that have the developing bones of a fetus preserved within the abdominal cavity, evidence of live birth; the fertilized egg was not laid in typical reptilian fashion but was retained within the body of the female until the fetus was fully developed.

Ichthyosaurs have been collected from many localities, but Berlin-Ichthyosaur State Park has the largest concentration of the largest ani-

248

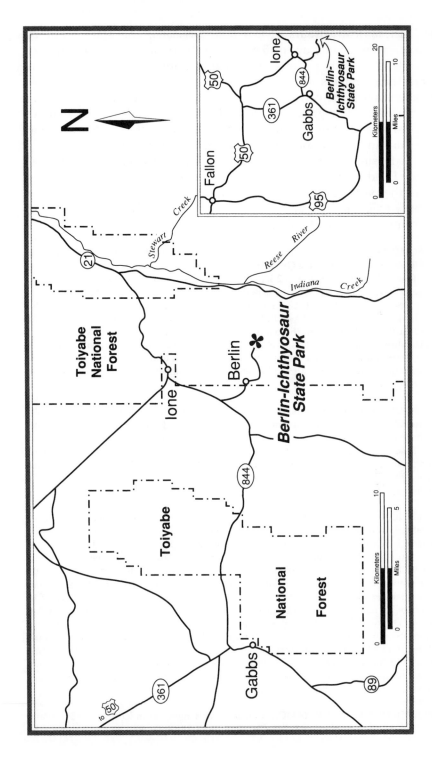

Figure 82. Location of Berlin-Ichthyosaur State Park, Nevada.

mals. These ichthyosaurs—*Shonisaurus popularis*, the state fossil of Nevada—were giants, up to 15 meters (50 feet) long. The accumulation of fossils is a death assemblage: individual animals died in the open ocean; their remains sank, some in the shallow waters surrounding offshore islands; the carcasses were oriented by the tides and currents, and buried by fine-grained muds. The ancient muds of the shallow marine shelf and stranded ichthyosaurs are now exposed in the Shoshone Mountains in the interior highlands of the Cordillera.

The fossils at Berlin-Ichthyosaur State Park, a United States National Natural Landmark, are known from a number of different quarry sites within the park, one of which is featured at an *in situ* exhibit. A large masonry relief model of an ichthyosaur on the up-hill side of the parking area complements the fossil displays.

DIRECTIONS: Access to Berlin-Ichthyosaur State Park is most direct and most reliable via State Route 361 and the town of Gabbs (Figure 82). Signs to the park are easily seen. Follow State Route 361 north of Gabbs 3.2 kilometers (2 miles) to State Route 844 and proceed eastward 33.5 kilometers (20.8 miles) to Berlin. The road is paved for the first 21.3 kilometers (13.2 miles), at 28 kilometers (17.3 miles) follow the left fork to Berlin. The parking area for the ichthyosaur exhibit is at the end of the road 3.7 kilometers (2.3 miles) beyond Berlin. Note that the road leading to the parking area is narrow and the last 0.8 kilometer (0.5 mile) has been graded out of the side of the canyon. Larger vehicles should consider parking at the campground below the ichthyosaur exhibit. Access to the park from the north via Austin or the south via Tonopah is possible but not recommended because the roads cross open range land, are not paved, and are not usually marked.

PUBLIC USE: Season and hours: The Berlin-Ichthyosaur State Park is open to the public year round and access to the outdoor exhibits is unrestricted. Access to the shelter enclosing the fossil exhibit is by way of guided tours given by park rangers. **Fees:** None to the park. **Food service:** Restaurants and stores are available at Gabbs. **Recreational activities:** Hiking trails featuring nature and history are found throughout the park. Camping (for which a fee is charged) and picnicking are also available. **Restrictions:** Collecting of fossils is prohibited.

EDUCATIONAL FACILITIES: Visitor Center: A building erected to house an *in situ* fossil exhibit serves some of the functions of a visitor center. Access to it and the enclosed fossils, however, is by way of scheduled guided tours. Nonetheless, interpretive facilities are accessible even when the building is not open to the public. Windows have been installed to allow visitors to see the interior exhibit. There are outdoor interpretive panels at both ends of the shelter, and there is an open-air exhibit of an ichthyosaur pelvic and tail section *in situ*. **Visitor Center hours:** The visitor center is open and tours are scheduled March 15 to November 15: daily; 10:00 A.M. and 2:00 P.M.; an additional tour scheduled Memorial Day to Labor Day at 4:00 P.M.. Tours may not always be available during the off-season, and visitors are advised to call to confirm tours. **Fees:** Ichthyosaur fossil tours: Adults $1.00, children 6–12 $0.50. **Staff programs:** Additional programs, such as special talks by park rangers and the Junior Paleontologist Program for 6–12 year olds, and special tours may be available in the summer. Schedules are posted at the Park Headquarters at Berlin and at the road junction 3.2 kilometers (2 miles) north of Gabbs.

FOR ADDITIONAL INFORMATION: Contact: Park Ranger, Berlin-Ichthyosaur State Park, Rt. 1, Box 32, Austin, Nevada 89310, (702) 964–2440. Information, especially during the off-season, can also be obtained from Nevada State Division of Parks, District Office, (702) 867–3001. **Read:** (1) Camp, Charles L. 1980. Large ichthyosaurs from the Upper Triassic of Nevada. Palaeontographica, Abteilung A: Palaeozoologie-Stratigraphie, volume 170, number 4–6, pp. 139–200. (2) Kosch, Bradley F. 1990. A revision of the skeletal reconstruction of *Shonisaurus popularis* (Reptilia: Ichthyosauria). Journal of Vertebrate Paleontology, volume 10, number 4, pp. 512–514. (3) Massare, Judy A. 1988. Swimming capabilities of Mesozoic marine reptiles: implications for method of predation. Paleobiology, volume 14, number 2, pp. 187–205.

42. Stewart Valley Paleontological Area

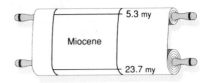

5.3 my

Miocene

23.7 my

Hawthorne, Nevada

The modern configuration of western North America took shape about 20 million years ago, in Miocene time. The Basin and Range topography that characterizes Nevada today was just beginning to develop. Volcanoes were active, producing extensive lava flows and great volumes of ash. A complex pattern of drainage and deposition developed as streams flowed down from the nascent ranges into growing basins, eroding and redepositing volcanic debris along with their sedimentary bedload. Drainage in the basins was not closed as it is today, the streams coalescing eventually to empty into the Pacific Ocean. Nor were the mountains of the Sierra Nevada to the west elevated yet. Thus no rainshadow existed in the Basin and Range; warm, moist air from the Pacific supplied abundant rainfall to the area.

The sediments in the Stewart Valley Paleontological Area are lake and river deposits layered with 30 meters (100 feet) of volcanic ash. The ash lends itself well to radiometric dating and facilitates correlation even if the sediments are very different between sites. Two exposures yield a wide variety of plant and animal fossils.

At Muller's Amphitheater, volcanic debris was deposited in water; it settled to the bottom of a pond, gently burying leaves, fruits, pollen, various insects, and fish. The insect fauna is the best of its age anywhere in the world. The plant remains are classified to more than 60 different species and collectively called the Fingerrock Flora. They confirm a temperate deciduous forest of elms, oaks, maples, and walnuts with pines and sequoias along the southwestern coast of North America and indicate a warm, temperate climate with 90 to 100 centimeters (35 to 40 inches) of precipitation a year. The forest is almost identical to the Miocene Russell Forests preserved at Ginkgo Petrified Forest State Park [Site 6], the environmental conditions similar; the way in which the fossils are preserved is reminiscent of conditions at Florissant Fossil Beds National Monument [Site 22].

The Snail Cliff section is younger than that at Muller's Amphitheater,

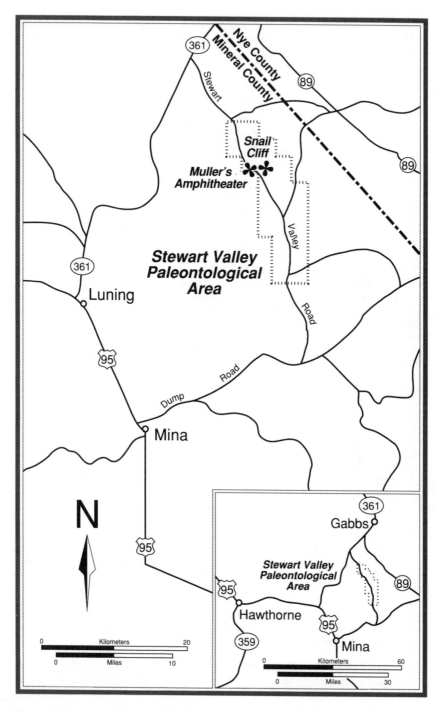

Figure 83. Location of Stewart Valley Paleontological Area, Nevada.

perhaps by as much as one million years. It exposes a tiny delta that formed where a stream emptied into a small lake. A variety in invertebrates, including snails, clams, and seed shrimp (ostracods) that lived in the deltaic sediments and on the lake bottom are preserved *in situ*. Fossils of large vertebrates have also been found, confirming that horses, camels, rhinoceros, and bear lived on the shores of the lake.

Fossils can be seen at each site. The sediments on the Snail Cliff face are distinctly layered, and the snails and clams can be seen weathering out of the soft earth. The hills at Muller's Amphitheater are made of exposed tuff, and the fossils, plant remains in particular, can be seen by examining the litter of platy fragments at the bases of the hills. The Stewart Valley Paleontological Area is a valuable scientific locality; the fossils and the sediments in which they occur are particularly fragile and have been designated by the Bureau of Land Management as an Area of Critical Environmental Concern.

DIRECTIONS: Access to the Stewart Valley Paleontological Area from the north is via State Route 361: 4.2 kilometers (2.6 miles) southwest of the Nye-Mineral counties line the unnamed Stewart Valley Road leaves the highway at Rawhide Ranch (Figure 83). Access from the south is via United States Route 95: 0.8 kilometer (0.5 mile) north of the settlement of Mina a paved road called Dump Road leaves the highway, beyond the dump the unnamed Stewart Valley road is unpaved but well-maintained. Signs marking entrance to the Stewart Valley Paleontological Area are encountered 11.6 kilometers (7.2 miles) south of State Route 361 and 33.5 kilometers (20.8 miles) north of United States Route 95. Note that the sites themselves and the signs that mark them are not prominent. The Snail Cliff site, on the east side of the road and marked by an interpretive panel, is 19.9 kilometers (12.3 miles) south of State Route 361 and 38.9 kilometers (24.1 miles) north of United States Route 95. The Muller's Amphitheater site is also on the east side of the road a distance of 0.7 kilometer (0.4 mile) south of the Snail Cliff site and is marked by a restricted side road and sign cautioning against removal of fossils and vehicular use of road. The walk to Muller's Amphitheater is about 0.8 kilometer (0.5 mile) along the restricted road up the draw.

PUBLIC USE: Season and hours: The Stewart Valley Paleontological Area is on public lands administered by the Bureau of Land Management and is open to the public year round. It is, however, a very fragile area, and the Bureau of Land Management is particularly concerned about protecting the integrity of the sites. To that end, any unauthorized collecting of fossils is strictly prohibited, and any violation will be prosecuted to the full extent of the law. Access by vehicles is limited to those roads and trails which are posted as open. Two sites, Muller's Amphitheater and Snail Cliff, are open to the public and offer limited interpretation. **Fees:** None. **Recreational activities:** None on site. **Restrictions:** Collecting of fossils is prohibited.

EDUCATIONAL FACILITIES: Interpretive sign: There is an interpretive panel at the Snail Cliff site, another at the restricted road that leads to Muller's Amphitheater. Each describes the fossils and the environments in which they lived.

FOR ADDITIONAL INFORMATION: Contact: Bureau of Land Management, Walker Resource Area Manager, Carson City District Office, 1535 Hot Springs Road, Suite 300, Carson City, Nevada 89706, (702) 882–1631.

43. Hancock Park

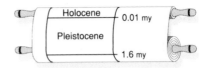

Holocene	0.01 my
Pleistocene	
	1.6 my

Los Angeles, California

Along the Pacific coast in southern California, for the last 40,000 years, oil has migrated from natural reservoirs in rocks upward along cracks that developed within the crust of the earth. The crude oil accumulated in shallow pools on the surface as a dense, sticky tar—natural asphalt. Today the tar pits are an important store of fossils.

The most famous deposit of tar-entrapped flora and fauna in North America is at Rancho La Brea; part of it is a site partially preserved in Hancock Park. The asphalt in pools on the surface formed natural traps for plants and animals. The animals ventured onto the pools, perhaps searching for food and water or being chased by predators, and became mired. Unable to extricate themselves, they died; their soft parts decayed, the bones became disarticulated. With time, the bones became impregnated with asphalt. The bones in some deposits are mixed and abraded, having been rubbed one against the other by the gentle churning induced by natural gas bubbling upward through the asphalt.

More than 560 species of animals and plants have been identified from the asphalt deposits: mammals, birds, reptiles, amphibians, and fish; freshwater and terrestrial invertebrates; pollen, leaves, and seeds. The mammal assemblage alone is represented by at least 60 species, but the composition is unusual. Remains of carnivores are by far the most numerous fossils: the best known is the saber-toothed cat, *Smilodon californicus*, the state fossil of California; among the largest is the American lion, *Panthera atrox*; puma, bobcat, jaguar and coyote are found; but the most common is the extinct dire wolf, *Canis dirus*.

Carnivores, the least numerous animals in an ecosystem and usually the least numerous in a fossil assemblage, constitute about 90 percent of the mammalian fauna at Rancho La Brea. Such a high proportion of carnivores suggests that the animals were attracted to the dead and dying animals already mired in the asphalt. Each carnivore was then itself trapped and acted as fresh bait. It was a death trap, and in that feature, it is not unlike Cleveland-Lloyd Dinosaur Quarry [Site 39].

Bones of herbivores are much less common; it seems the herbivores only occasionally and accidentally became trapped. Nonetheless hundreds of thousands of bones have been recovered and many species are present: two extinct species of bison, two horses, two camels, mammoth and mastodon, deer and antelope, and three ground sloths. Rabbits, rodents, insectivores, and bats: all are represented.

The number of birds preserved is also high; 135 species have been identified. It appears, that like the carnivorous mammals, the carnivorous or scavenging eagles, hawks, falcons, condors, and vultures were attracted by mired prey and became trapped themselves. Bird bones are delicate and easily destroyed, but in the asphalt they were protected.

The fossil flora represents the distinct plant associations that occupied the different habitats in the area. The 160 species that are known characterize four main plant associations. Reconstruction shows that chaparral forests covered the high country, and redwoods were restricted to sheltered canyons; the streams were lined with deciduous trees, the coastal slope supported sagebrush and groves of cypress and pine.

The first scientific excavations at Rancho La Brea were conducted in 1901, and the value of the fossils was quickly recognized. The Natural History Museum of Los Angeles County, which administers the site, first excavated in 1913–1915. At present, more than one million bones and a total of four million fossils from the asphalt deposits are housed in the George C. Page Museum. The unique fossil resource is recognized by the designation of the area as a United States National Natural Landmark.

DIRECTIONS: Hancock Park and the George C. Page Museum is located within the city of Los Angeles at 5801 Wilshire Boulevard two blocks east of Fairfax Avenue (Figure 84). It is bounded by Sixth Street on the north, Curson Street on the east, and Ogden Street on the west. Parking is off Curson Street. A good street map of the city is indispensable for visitors intending to drive. Access may be also gained by public transportation.

PUBLIC USE: Season and hours: In Hancock Park, open air and enclosed tar pits are accessible to the public. One can see partially excavated deposits, bubbling asphalt, an excavation in progress (during July and August), and models of extinct animals erected around the tar pits in the park. The park and the outdoor exhibits are open to the public year round. **Fees:** None. **Food service:** There is a snack bar adjacent to the museum, and restaurants and stores are available nearby. **Recreational activities:** None on site. **Handicapped facilities:** All public use areas are accessible by wheelchair. **Restrictions:** Collecting of fossils is prohibited.

EDUCATIONAL FACILITIES: Museum: The George C. Page Museum of La Brea Discoveries, located in Hancock Park, is dedicated to the preservation and interpretation of Ice Age fossils found in the tar pits of Rancho La Brea. It has a variety of exhibits of Rancho La Brea fossils including numerous skeletons, multi-media presentations, and hands-on learning (for example, to illustrate the stickiness of asphalt). Of particular interest is the paleontology laboratory, a semi-circular glass-enclosed area, which allows visitors to observe fossil preparation in progress. This is a working laboratory where scientists and volunteers work on fossils daily (except Tuesdays). **Museum hours:** The Page Museum is open year round: Tuesday to Sunday; from 10:00 A.M. to 5:00 P.M.; except New Year's Day

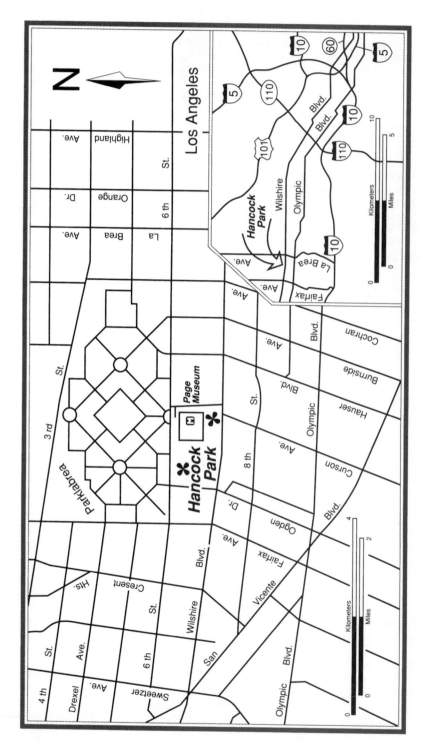

Figure 84. Location of Hancock Park, Los Angeles, California.

(January 1), Thanksgiving Day, and Christmas Day (December 25); open some school holiday Mondays (confirm by telephone). **Fees:** Adults $3.00, students and senior citizens $1.50, children aged 5–12 years $0.75 (free on second Tuesday of each month). **Bookstore:** The Page Museum Shop is a gift shop that sells a wide range of gift items and educational material. It stocks a variety of books including many on the local discoveries. **Tour guide:** Tours of the museum are available year round: Saturday and Sunday, 11:30 A.M.; Tuesday to Sunday, 2:00 P.M.. Tours of the Rancho La Brea Tar Pits are available year round: Wednesday to Sunday, 1:00 P.M. School groups and organized groups may arrange tours by reservation, phone (213) 857-6306. **Interpretive sign:** Interpretive signs are posted at exhibits within the park.

FOR ADDITIONAL INFORMATION: Contact: George C. Page Museum, 5801 Wilshire Boulevard, Los Angeles, California 90036, (213) 936–2230 or (213) 857–6311. **Read:** (1) Harris, John M., and George T. Jefferson (editors). 1985. Rancho La Brea: Treasures of the Tar Pits. Natural History Museum of Los Angeles County, Science Series 31. (2) Stock, Chester, 1956. Rancho La Brea: A Record of Pleistocene Life in California. Sixth Edition. Natural History Museum of Los Angeles County, Science Series 20, Paleontology Number 11.

44. Ralph B. Clark Regional Park

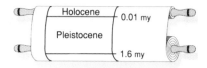

Buena Park, California

The fossils at Ralph B. Clark Regional Park document the geological events that dramatically altered the topography of southern California over the last five million years. The history begins in Pliocene time when a small but deep ocean basin, the Los Angeles Basin, formed along the margin of the continent. Its depth at any one time is estimated to have been approximately 1500 meters (5000 feet). The basin was characterized by subsidence for about three million years; simultaneously, sediments derived from the continent were deposited in it, accumulating to a thickness of more than 3500 meters (11,500 feet). These deep water sediments contain marine fossils of foraminifera, mollusks, and echinoderms.

The way in which tectonic processes were operating on the Los Angeles Basin changed about two million years ago. Subsidence stopped; uplift began. Over time, the basin was gradually uplifted; shallow marine conditions came to prevail, and eventually the Los Angeles Basin was subaerially exposed.

Early in the Pleistocene when the Los Angeles Basin was still relatively deep, the area now represented in the park was submarine; marine sediments were being deposited entombing among the invertebrate fossils the bones of whales, birds, sea lions, dolphins, and fish. In more recent Pleistocene time, when uplift and faulting had turned the marine basin into a grassy coastal plain, streams flowed across the new land surface and deposited plant and animal remains along with muds, sands, and gravels. Grassy meadows, oak woodlands, and oxbow lakes along a meandering river provided a wide range of habitats for plants and animals: animals such as mammoths and ground sloths, ringtail cats and wolves, horses and tapirs, bison and llama. Freshwater fish and mollusks represent the life in the river.

DIRECTIONS: The Ralph B. Clark Regional Park is located within the greater Los Angeles area in the cities of Fullerton and Buena Park (Figure 85). Access is via Beach Boulevard (from either Riverside or Santa Ana freeways) and then east on Rosecrans Avenue about 0.8 kilometer (0.5 mile). The park straddles Rosecrans Avenue.

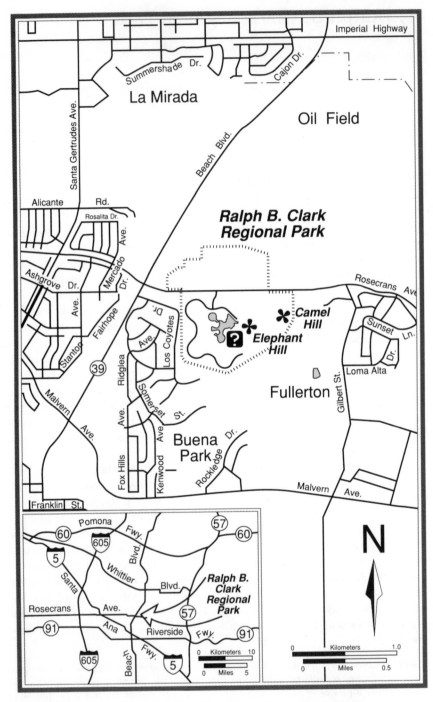

Figure 85. Location of Ralph B. Clark Regional Park, Buena Park, California.

PUBLIC USE: Season and hours: The Ralph B. Clark Regional Park is open year round: daily; from 7:00 A.M. to 6:00 P.M.; with extended hours April 1 to October 31 to 9:00 P.M.. **Fees:** A parking fee of $2.00 daily/vehicle ($10.00/bus) is charged; annual day-use passes are also available. **Food service:** Restaurants and stores are available in the area. **Recreational activities:** Hiking areas, horseshoe pits, a bicycle trail, tennis and volleyball courts, and softball and baseball fields are available as well as a picnic and a playground area. Fishing is also permitted. **Handicapped facilities:** Many areas of the park, including the visitor center, are accessible by wheelchair. **Restrictions:** Collecting of fossils is prohibited.

EDUCATIONAL FACILITIES: Visitor Center: An interpretive center features fossil exhibits, a glass-enclosed laboratory for observation, and visitor-activated exhibits. Park staff actively work at fossil-bearing sites within the park, and visitors may have the opportunity to observe the process. **Visitor Center hours:** The visitor center is open year round: Tuesday through Sunday; from 9:00 A.M. to 3:00 P.M.. **Fees:** Included in the entrance fees listed above. **Trails:** The Ralph B. Clark Park features outdoor excavation areas where sediments of different ages are exposed: at Camel Hill and Elephant Hill a 70,000 year old deposit is exposed; a 1.4 million year old marine unit is exposed north of Rosecrans Ave. **Tour guide:** Guided tours that include both laboratory and field exploration for groups of 15 people or more can be arranged (15 days advance notice is required). If quarries are being excavated, individual tours to the digs can be arranged. A fee of $1.00/person/hour is charged. **Staff programs:** Accepting and training volunteers is one of the park's educational objectives. **Note:** Field weeks are being planned during which time the emphasis in the park will be on public participation in the fossil collecting process, and people interested in learning paleontological techniques can participate as volunteers. The park runs a free-of-cost program for mentally and physically disadvantaged people who would like some experience working with fossils.

FOR ADDITIONAL INFORMATION: Contact: Paleontologist, Ralph B. Clark Regional Park, 8800 Rosecrans Avenue, Buena Park, California 90621, (714) 670–8052.

45. Anza-Borrego Desert State Park

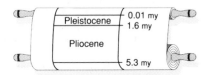

Borrego Springs, California

The San Andreas Fault is a relatively young geological structure, a deep fracture in the crust of North America that began to develop some 25 million years ago in latest Oligocene or earliest Miocene time. With time, the fracture opened, and the Gulf of California and a deep ocean basin known as the Salton Trough formed. The Pacific Ocean flooded the Salton Trough; an inland sea extended northward, its remnant the present Salton Sea. The ancestral Colorado River emptied into the sea; the depth of the Salton Trough and the amount of sediment supplied to it can be gauged from the 4000 meter (13,000 foot) thick sedimentary sequence known in Anza-Borrego Desert State Park. When subsidence ceased, the Salton Trough filled with sediment, and the Gulf of California retreated southward. Where once was open sea, now the river meandered across its floodplain, eventually to deposit its load of sediment as a delta in a remnant lake.

The fossils at Anza-Borrego Desert State Park document both phases of the geological history of the Salton Trough. In either case, in the not-so-distant past, the climates and environments were totally unlike those of the present. The remnants of a reef made up of oysters, clams, and corals can be seen in the exposures of the Imperial Formation, such as at Fishers Creek. The animals whose remains make up the reef lived in a shallow, open sea during Miocene time. Eight million years later the sea had retreated; in the deltaic and river sediments that overlie the marine deposits are the remains of savannah plants and animals that are Late Pliocene and Pleistocene in age: horses, camels, mammoths, and saber-toothed cats; rabbits, rodents, and shrews. Desert conditions prevail today but the rainshadow cast by young mountains did not always exist. The sediments are exposed in the badlands at Fonts Point.

DIRECTIONS: Access to Anza-Borrego Desert State Park is via State Route 78, which crosses approximately through the center of the park (from the east and Interstate 10 take State Route 78 or 86; from the west via State Route 78 or 79); access from Ocotillo on Interstate 8 is along a paved secondary route, S–2, into and across the park (Figure 86).

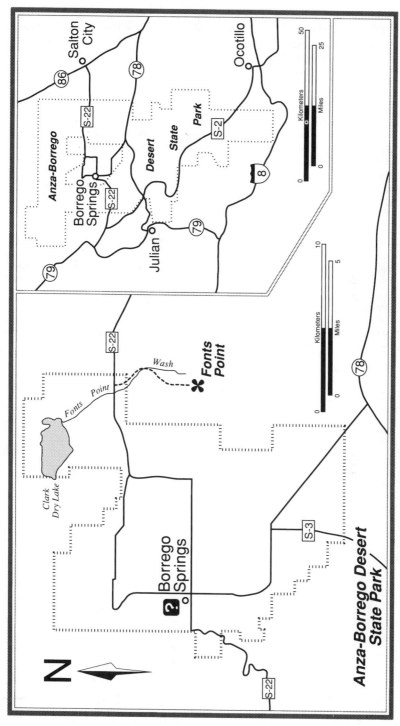

Figure 86. Location of Anza-Borrego Desert State Park, California. Note that the town of Borrego Springs and surrounding area within the park are private property.

263

Access to the Fonts Point fossil area is via route S–22 east of Borrego Springs, and then south for 6.4 kilometers (4 miles) on a partly graded road that follows Fonts Point Wash. The road widens to accommodate parking, from which point a trail 100 meters (300 feet) long leads to a view point, then turns abruptly to the right and loops back to join Fonts Point road. Park rangers provide directions to the fossil exposures at Fishers Creek.

PUBLIC USE: Season and hours: Anza-Borrego Desert State Park is open to the public year round, but portions such as Coyote Canyon are subject to seasonal closure (June 16–September 16) to allow bighorn sheep undisturbed access to waterholes. The park offers a variety of visitor activities during the winter, the primary visitor season. **Fees:** $3.00 daily/vehicle, $1.00 daily/dog, and $30.00 daily/bus. A senior citizen discount ($1.00) is available. **Food service:** Restaurants and stores are available at Borrego Springs. **Recreational activities:** Camping (for which an additional fee is charged with a discount available for senior citizens) including primitive camping, hiking, backpacking, and horseback riding are available. **Restrictions:** Collecting of fossils is prohibited.

EDUCATIONAL FACILITIES: Visitor Center: The visitor center, located in Borrego Springs (on Palm Canyon Drive) highlights the natural history of the desert and features a fossil room or laboratory. Park rangers offer tours of the Stout Paleontology Laboratory. **Visitor Center hours:** The visitor center is open year round: daily October to May from 9:00 A.M. to 5:00 P.M.; weekends and holidays June to September from 10:00 A.M. to 3:00 P.M. **Fees:** None. **Bookstore:** The bookstore features some books and items of local interest. **Interpretive sign:** The interpretive panels at Fonts Point are accessible year round.

FOR ADDITIONAL INFORMATION: Contact: District Superintendent, Anza-Borrego Desert State Park, P. O. Box 299, Borrego Springs, California 92004, (619) 767–5311.

*Ashfall Fossil Beds State Historical Park

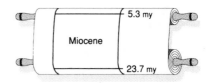

5.3 my

Miocene

23.7 my

Royal, Nebraska

Ashfall Fossil Beds State Historical Park is a new fossil site developed for public access (open June 1, 1991). The focus of the park is the excavation of Miocene vertebrates, 10 million year old fossils preserved in volcanic ash. There will be an interpretive center with displays and a visible preparation lab. A large building called a rhino lab has been built over the fossil quarry where paleontologists will excavate and leave *in situ* additional skeletons. The park is located approximately 160 kilometers (100 miles) west of Sioux City, Iowa, on United States Route 20 between the towns of Royal and Orchard in Nebraska: 3.2 kilometers (2 miles) west and 9.6 kilometers (6 miles) north of Royal; 6.4 kilometers (4 miles) east and 9.6 kilometers (6 miles) north of Orchard. The fossil site will be open Memorial Day to Labor Day: daily; from 9:00 A.M. to 5:00 P.M. A nominal fee (or a valid Nebraska State Park Permit) will be charged.

Section III

Sources of
Additional Information

Museum Exhibits

Regional museums in western Canada and United States feature exhibits in paleontology. The exhibits usually interpret the fossil history of the province or state in which they are located. The following list of regional museums does not include the exhibits located at many of the fossil sites. Detailed information on these and other museums is available in the most recent edition of the American Association of Museums' *The Official Museum Directory* (published annually by National Register Publishing Company, Inc., 3004 Glenview Road, Wilmette, Illinois 60091) and *Official Directory of Canadian Museums and Related Institutions* (published annually by the Canadian Museums Association, 280 Metcalfe Avenue, Suite 400, Ottawa, Ontario K2P 1R7).

Canada

Alberta

Provincial Museum of Alberta, 12845 102 Avenue, Edmonton, Alberta, T5N 0M6.
Royal Tyrrell Museum of Palaeontology, Box 7500, Drumheller, Alberta, T0J 0Y0.

Manitoba

Manitoba Museum of Man and Nature, 190 Rupert Avenue, Winnipeg, Manitoba, R3B 0N2.

Saskatchewan

Saskatchewan Museum of Natural History, Wascana Park, Regina, Saskatchewan, S4P 3V7.

United States

Arizona

Museum of Northern Arizona, Fort Valley Road, Flagstaff, Arizona 86001.
Arizona State Museum, University of Arizona, Tucson, Arizona 85721.

California

George C. Page Museum of La Brea Discoveries, 5801 Wilshire Boulevard, Los Angeles, California 90036.
Museum of Paleontology, Earth Sciences Building, University of California, Berkeley, California 94720.

Los Angeles County Natural History Museum, 900 Exposition Boulevard, Los Angeles, California 90007.

Colorado
Denver Museum of Natural History, City Park, Denver, Colorado 80205.

Dinosaur Valley, Museum of Western Colorado, Fourth and Main Streets, P.O. Box 20000–5020, Grand Junction, Colorado 81501–5020.

Museum of Geology, University of Colorado, Boulder, Colorado 80309.

Idaho
Museum of Natural History, Idaho State University, Pocatello, Idaho 83209.

Kansas
Museum of Natural History, University of Kansas, Lawrence, Kansas 66045.

Sternberg Memorial Museum, Fort Hays State Museum, Hays, Kansas 67601.

Montana
Museum of the Rockies, Montana State University, Bozeman, Montana 59717.

Nebraska
Earth Science Museum, Chadron State College, Chadron, Nebraska 69337.

State Museum, University of Nebraska, Lincoln, Nebraska 68588.

Trailside Museum (University of Nebraska), Fort Robinson State Park, Crawford, Nebraska 69339.

Nevada
Nevada Geology Museum, University of Nevada, Las Vagas, Nevada 89154.

The Nevada State Museum, 600 North Carson Street, Carson City, Nevada 89710.

New Mexico
New Mexico Natural History Museum, 1801 Mountain Road, N.W., Albuquerque, New Mexico 87194.

Oklahoma
Stovall Museum, Oklahoma Museum of Natural History, University of Oklahoma, Norman, Oklahoma 73019.

Oregon
Geology Museum, University of Oregon, Eugene, Oregon 97403.

South Dakota
Museum of Geology, South Dakota School of Mines and Technology, Rapid City, South Dakota 57701.

Texas
Dallas Museum of Natural History/Dallas Aquarium, Fair Park Station, Dallas, Texas 75226.

Houston Museum of Natural History, 1 Hermann Circle Drive, Hermann Park, Houston, Texas 77030.

Panhandle Plains Museum, 2401 Fourth Avenue, Canyon, Texas 79015.

Petroleum Museum/Permian Reef Geology Museum, 1500 Interstate 20 West, Midland, Texas 79701.

Texas Memorial Museum, 2400 Trinity, Austin, Texas 78705.

Utah

Utah Field House of Natural History, Vernal, Utah 84078.

Museum of Natural History, University of Utah, Salt Lake City, Utah 84112.

Prehistory Museum, College of Eastern Utah, Price, Utah.

Washington

Thomas Burke Memorial Washington State Museum, University of Washington, Seattle, Washington 98105.

Wyoming

Geological Museum, University of Wyoming, Laramie, Wyoming 82070.

Publications

There is a vast body of literature, both professional and popular, on paleontology and related disciplines. The list of books, articles, and journals that follows represents a selected bibliography, the titles chosen such that both popular and technical publications on a wide range in topics are represented. Some are broad in scope, others focus on specific concepts, fossils, or geological time and place. A complete list of the publications referred to in the site descriptions is also included.

Young Readers

Archer, Jules. 1976. From Whales to Dinosaurs. New York, New York: St. Martin's Press.

Arthur, Alex. 1989. Shell. Toronto, Ontario: Eyewitness Books, Stoddard Publishing Company Ltd.

Halstead, L. Beverly. 1975. The Ecology and Evolution of the Dinosaurs. London, England: Peter Lowe.

Halstead, L. Beverly. 1978. The Evolution of the Mammals. London, England: Peter Lowe.

Halstead, L. Beverly. 1982. The Search for the Past: Rocks, Fossils, Tracks, and Trails; the Search for the Origins of Life. Garden City, New York: Doubleday and Company, Inc.

Halstead, L. Beverly, and Jenny Halstead. 1987. Dinosaurs. New York, New York: Sterling Publishing Company.

Hublin, Jean-Jacques. 1984. The Hamlyn Encyclopedia of Prehistoric Animals. London, England; New York, New York: Hamlyn.

Lambert, David, and the Diagram Group. 1983. A Field Guide to Dinosaurs. New York, New York: Avon Books.

Lambert, David, and the Diagram Group. 1985. The Field Guide to Prehistoric Life. New York, New York: Facts on File, Inc.

Lambert, David, and the Diagram Group. 1988. The Field Guide to Geology. New York, New York: Facts on File, Inc.

Lauber, Patricia. 1987. Dinosaurs Walked Here: And Other Stories Fossils Tell. New York, New York: Bradbury Press.

Parker, Steve. 1988. Skeleton. New York, New York: Eyewitness Books, Alfred A. Knopf.

Sattler, Helen Roney. 1981. Dinosaurs of North America. New York, New York: Lothrop, Lee and Shepard Books.

Taylor, Paul D. 1990. Fossil. Toronto, Ontario: Eyewitness Books, Stoddard Publishing Company Ltd.

Historical Perspectives

Historical Writings

Darwin, Charles. 1988. Diary of the Voyage of H. M. S. Beagle. Edited by Richard Darwin Keynes. Cambridge, England; New York, New York: Cambridge University Press.

Hitchcock, Eduard. 1848. An attempt to discriminate and describe the animals that make the fossil footmarks of the United States and especially New England. Memoir of the American Academy of Arts and Science, series 2, volume 3, pp. 129–256.

Hutton, James. 1788. Theory of the Earth. Transactions of the Royal Society of Edinburgh 1, pp. 209–305.

Osborn, Henry Fairfield. 1918. The Origin and Evolution of Life: On the Theory of Action, Reaction, and Interaction of Energy. New York, New York: Charles Scribner's Sons.

Playfair, John. 1802. Illustrations of the Huttonian Theory of the Earth. Edinburgh, Scotland: William Creech.

History of Paleontology and Biographies

Andrews, Henry N. 1980. The Fossil Hunters: In Search of Ancient Plants. Ithaca, New York; London, England: Cornell University Press.

Boorstin, Daniel J. 1983. The Discoverers: A History of Man's Search to Known His World and Himself. New York, New York: Random House.

Brent, Peter. 1981. Charles Darwin: A Man of Enlarged Curiosity. London, England: William Heineman Limited.

Buffetaut, Eric. 1987. A Short History of Vertebrate Palaeontology. London, England; Wolfeboro, New Hampshire: Croom Helm.

Colbert, Edwin H. 1989. Digging into the Past: An Autobiography. New York, New York: Dembner Books.

Desmond, Adrian J. 1982. Archetypes and Ancestors: Paleontology in Victorian London, 1850–1875. Chicago, Illinois: The University of Chicago Press.

Faul, Henry, and Carol Faul. 1983. It Began With A Stone: A History of Geology from the Stone Age to the Age of Plate Tectonics. New York, New York: John Wiley and Sons.

Gregory, Joseph T. 1979. North American vertebrate paleontology, 1776–1976. Pp. 305–335 in Cecil J. Scheer (editor), Two Hundred Years of Geology in America. Proceedings of the New Hampshire Bicentennial Conference on the History of Geology. Hanover, New Hampshire: The University Press of New England.

Howard, Robert West. 1975. The Dawnseekers: The First History of American Paleontology. New York, New York; London, England: Harcourt, Brace, Jovanovich.

Hull, David L. 1973. Darwin and his Critics: The Reception of Darwin's Theory of Evolution by the Scientific Community. Cambridge, Massachusetts: Harvard University Press. Reprinted 1983, Chicago, Illinois: The University of Chicago Press.

Lanham, Urless N. 1973. The Bone Hunters. New York, New York; London, England: Columbia University Press.

Marchant, James. 1975. Alfred Russel Wallace: Letters and Reminiscences. New York, New York: Arno Press.

Olson, Everett C. 1990. The Other Side of the Medal: A Paleobiologist Reflects on the Art and Serendipity of Science. Blacksburg, Virginia: The McDonald and Woodward Publishing Company.

Plate, Robert. 1964. The Dinosaur Hunters: Othniel C. Marsh and Edward D. Cope. New York, New York: David McKay Company Inc.

Rudwick, Martin J. S. 1972. The Meaning of Fossils: Episodes in the History of Palaeontology. London, England: Macdonald and Co. Ltd.; New York, New York: American Elsevier Publishing Co. Inc.

Russell, Loris S. 1966. Dinosaur hunting in western Canada. Royal Ontario Museum, Life Sciences Contributions, Number 70.

Shor, Elizabeth Noble. 1971. Fossils and Flies: The Life of A Compleat Scientist, Samuel Wendell Williston (1851–1918). Norman, Oklahoma: University of Oklahoma Press.

Shor, Elizabeth Noble. 1974. The Fossil Feud: Between E. D. Cope and O. C. Marsh. Hicksville, New York: Exposition Press.

Simpson, George Gaylord. 1978. Concession to the Improbable: An Unconventional Autobiography. New Haven, Connecticut; London, England: Yale University Press.

Context and Overview

There are many publications, both popular and technical, which provide syntheses of current knowledge in the various disciplines in paleontology. The list below includes titles aimed at readers interested in general information, as well as titles for those interested in more detailed study.

Concepts

Berry, William B. N. 1987. Growth of a Prehistoric Time Scale: Based on Organic Evolution. Revised Edition. Palo Alto, California: Blackwell Scientific Publications.

Darwin, Charles. 1859. On the Origin of Species by Means of Natural Selection, or the Preservation of Favoured Races in the Struggle for Life. London, England: J. Murray.

Fox, Richard C. Paleoscene #1. Species in paleontology. Geoscience Canada, volume 13, number 2, pp. 73–84.

Futuyma, Douglas J. 1983. Science on Trial: The Case for Evolution. New York, New York: Pantheon Books.

Gould, Steven Jay 1987. Time's Arrow, Time's Cycle: Myth and Metaphor in the Discovery of Geological Time. Cambridge, Massachusetts; London, England: Harvard University Press.

Harland, W. B. A., A. V. Cox, P. G. Llewellyn, C. A. G. Picton, A. G. Smith, and R. Walters. 1990. A Geologic Time Scale 1989. New York, New York: Cambridge University Press.

Lovelock, James E. 1987. Gaia: A New Look at Life on Earth. Oxford, England; New York, New York: Oxford University Press.

Lovelock, James E. 1988. The Ages of Gaia: A Biography of our Living Earth. New York, New York: W. W. Norton and Company.

Nitecki, Matthew H. (editor). 1989. Evolutionary Progress. Chicago, Illinois: The University of Chicago Press.

Ruse, Michael. 1979. The Darwinian Revolution: Science Red in Tooth and Claw. Chicago, Illinois: The University of Chicago Press.

Ruse, Michael. 1982. Darwinism Defended: A Guide to the Evolution Controversies. Reading, Massachusetts: Addison-Wesley.

Ruse, Michael (editor). 1988. But Is It Science? The Philosophical Question in the Creation/Evolution Controversy. Buffalo, New York: Prometheus Books.

Origins

Fabian, A. C. (editor). 1989. Origins. New York, New York: Cambridge University Press.

Cairns-Smith, A. G. 1985. Seven Clues to the Origin of Life. New York, New York: Cambridge University Press.

Glaessner, Martin F. 1984. The Dawn of Animal Life: A Biohistorical Study. Cambridge, England: Cambridge University Press (Cambridge Earth Science Series).

Margulis, Lynn. 1970. Origin in Eukaryotic Cells: Evidence and Research Implications for a Theory of the Origin and Evolution of Microbial, Plant, and Animal Cells on the Precambrian Earth. New Haven, Connecticut: Yale University Press.

Margulis, Lynn. 1981. Symbiosis in Cell Evolution: Life and Its Environment on the Early Earth. New York, New York: W. H. Freeman and Company.

Margulis, Lynn. 1984. Early Life. Boston, Massachusetts: Jones and Bartlett.

Margulis, Lynn, and Dorion Sagan. 1986a. Origins of Sex. New Haven, Connecticut: Yale University Press.

Margulis, Lynn, and Dorion Sagan. 1986b. Microcosmos: Four Billion Years of Evolution from our Microbial Ancestors. New York, New York: Summit Books.

Schopf, J. William (editor). 1983. Earth's Earliest Biosphere: Its Origin and Evolution. Princeton, New Jersey: Princeton University Press.

Vidal, Gonzalo. 1984. The oldest eukaryotic cells. Scientific American, volume 250, number 2, pp. 48–57.

Paleontology

Allen, Keith C., and Derek E. G. Briggs (editors). 1990. Evolution and the Fossil Record. Washington, DC: Smithsonian Institution Press.

Axelrod, Daniel I. 1988. An interpretation of high montane conifers in western Tertiary floras. Paleobiology, volume 14, number 3, pp. 301–306.

Bakker, Robert T. 1986. The Dinosaur Heresies: New Theories Unlocking the Mystery of the Dinosaurs and Their Extinction. New York, New York: William Morrow and Company, Inc.

Behrensmeyer, Anna K., and Andrew P. Hill (editors). 1980. Fossils in the Making: Vertebrate Taphonomy and Paleoecology. Chicago, Illinois: The University of Chicago Press.

Black, Rhona M. 1989. The Elements of Palaeontology. Second Edition. Cambridge, England; New York, New York: Cambridge University Press.

Boardman, Richard S., Alan H. Cheetham, and Albert J. Rowell (editors). 1987. Fossil Invertebrates. Palo Alto, California: Blackwell Scientific Publications.

Briggs, Derek E. G., and Peter R. Crowther. 1990. Paleobiology: A Synthesis. Palo Alta, California: Blackwell Scientific Publications.

Carpenter, Kenneth, and Philip J. Currie (editors). 1990. Dinosaur Systematics: Approaches and Perspectives. New York, New York: Cambridge University Press.

Carroll, Robert L. 1988. Vertebrate Paleontology and Evolution. New York, New York: W. H. Freeman and Company.

Clarkson, Euan N. K. 1986. Invertebrate Palaeontology and Evolution. Second Edition. London, England: George Allen and Unwin.

Case, Gerard R. 1982. A Pictorial Guide to Fossils. New York, New York; Toronto, Ontario: Van Nostrand Reinhold Company.

Cowen, Richard. 1990. History of Life. Boston, Massachusetts: Blackwell Scientific Publications.

Czerkas, Sylvia J., and Everett C. Olson. 1987. Dinosaurs Past and Present. Seattle, Washington; London, England: Natural History Museum of Los Angeles County in association with University of Washington Press. Two volumes.

Davis, Simon J. M. 1987. The Archaeology of Animals. New Haven, Connecticut: Yale University Press.

Desmond, Adrian J. 1975. The Hot Blooded Dinosaurs: A Revolution in Paleontology. London, England: Briggs and Blond.

Eldredge, Niles. 1987. Life Pulse: Episodes from the Story of the Fossil Record. New York, New York; Oxford, England: Facts on File Inc.

Eldredge, Niles, and Steven M. Stanley (editors). 1984. Living Fossils. New York, New York: Springer Verlag.

Farlow, James O. (editor). 1989. Paleobiology of the Dinosaurs. Boulder, Colorado: Geological Society of America (South-Central Section), Special Paper 238.

Feduccia, Alan. 1980. The Age of Birds. Cambridge, Massachusetts: Harvard University Press.

Gall, Jean-Claude. 1983. Ancient Sedimentary Environments and the Habitats of Living Organisms: Introduction to Palaeoecology. Berlin, West Germany: Springer-Verlag.

Genoways, Hugh H., and Mary R. Dawson (editors). 1984. Contributions in Quaternary Vertebrate Paleontology: A Volume in Memorial to John E. Guilday. Carnegie Museum of Natural History Special Publication Number 8.

Gould, Stephen Jay. 1980. The Panda's Thumb: More Reflections in Natural History. New York, New York: W. W. Norton and Company.

Gould, Stephen Jay. 1985. The Flamingo's Smile: Reflections in Natural History. New York, New York: W. W. Norton and Company.

Gould, Stephen Jay. 1991. Bully for Brontosaurus. New York, New York: W. W. Norton and Company.

Guthrie, R. Dale. 1990. Frozen Fauna of the Mammoth Steppe: The Story of Blue Babe. Chicago, Illinois: The University of Chicago Press.

Hopkins, David M., John V. Matthews, Jr., Charles E. Schweger, and Steven B. Young (editors). 1982. Paleoecology of Beringia. New York, New York: Academic Press.

Hotton, Nicholas, III, P. D. MacLean, J. J. Roth, and E. C. Roth (editors). 1986. The Ecology and Biology of Mammal-like Reptiles. Washington, DC: Smithsonian Institution Press.

Jeffries, R. P. S. 1986. The Ancestry of the Vertebrates. London, England: British Museum (Natural History).

Jones, O. A., and R. Endean. 1973. Biology and Geology of Coral Reefs. Volume 1: Geology 1. New York, New York; London, England: Academic Press.

Jones, O. A., and R. Endean. 1973. Biology and Geology of Coral Reefs. Volume 2: Biology 1. New York, New York; London, England: Academic Press.

Jones, O. A., and R. Endean. 1976. Biology and Geology of Coral Reefs. Volume 3: Biology 2. New York, New York; London, England: Academic Press.

Jones, O. A., and R. Endean. 1977. Biology and Geology of Coral Reefs. Volume 4: Geology 2. New York, New York; London, England: Academic Press.

Kurtén, Bjorn, and Elaine Anderson. 1980. Pleistocene Mammals of North America. New York, New York: Columbia University Press.

Levi-Setti, Riccardo. 1975. Trilobites: A Photographic Atlas. Chicago, Illinois; London, England: the University of Chicago Press.

Macdonald, James Reid. 1983. The Fossil Collector's Handbook: A Paleontology Field Guide. Englewood Cliffs, New Jersey: Prentice-Hall Inc.

McKerrow, W. S. (editor). 1978. The Ecology of Fossils: An Illustrated Guide. Cambridge, Massachusetts: The MIT Press.

Meyen, Sergei. 1987. Fundamentals of Palaeobotany. London, England: Chapman and Hall.

Moy-Thomas, J. A. 1971. Paleozoic Fishes. Second Edition, revised by R. S. Miles. London, England: Chapman and Hall, Ltd.

Nowlan, Godfrey S. 1986. Paleoscene Introduction. Paleontology: Ancient and modern. Geoscience Canada, volume 13, number 2, pp. 67–72.

Ostrom, John H., and John S. McIntosh. 1966. Marsh's Dinosaurs: The Collections from Como Bluff. New Haven, Connecticut; London, England: Yale University Press.

Padian, Kevin. 1986. Dawn of the Age of Dinosaurs: Faunal Change Across the Triassic-Jurassic Boundary. Cambridge, England; New York, New York: Cambridge University Press.

Paul, Gregory S. 1988. Predatory Dinosaurs of the World: A Complete Illustrated Guide. New York, New York; London, England: Simon and Schuster.

Paul, Gregory S. 1988. Physiological, migratorial, climatological, geophysical, survival, and evolutionary implications of Cretaceous polar dinosaurs. Journal of Paleontology, volume 62, number 4, pp. 640–652.

Porter, Stephen C. (editor). 1983. Late-Quaternary Environments of the United States. Volume 1. The Late Pleistocene. Minneapolis, Minnesota: University of Minnesota Press.

Radinsky, Leonard B. The Evolution of Vertebrate Design. Chicago, Illinois: The University of Chicago Press.

Raup, David M., and Steven M. Stanley. 1978. Principles of Paleontology. Second Edition. San Francisco, California: W. H. Freeman and Company.

Reader, John. 1986. The Rise of Life—The First 3.5 Billion Years. New York, New York: Knopf Inc.

Rixon, A. E. 1976. Fossil Animal Remains: Their Preparation and Conservation. London, England: The Athlone Press of the University of London.

Rudwick, Martin J. S. 1970. Living and Fossil Brachiopods. London, England: Hutchison and Company Ltd.

Ryland, J. S. 1970. Bryozoans. London, England: Hutchison and Company Ltd.

Schmid, Elizabeth. 1972. Atlas of Animal Bones: For Prehistorians, Archaeologists, and Quaternary Geologists. Amsterdam, Holland; New York, New York: Elsevier Publishing Company.

Schulte McMenamin, Mark A. S., and Dianna L. Schulte McMenamin. 1989. The Emergence of Animals: The Cambrian Breakthrough. New York, New York: Columbia University Press.

Shipman, Pat. 1981. Life History of a Fossil: An Introduction to Taphonomy and Paleoecology. Cambridge, Massachusetts: Harvard University Press.

Stewart, Wilson N. 1983. Paleobotany and the Evolution of Plants. Cambridge, England; New York, New York: Cambridge University Press.

Taylor, Thomas N. 1981. Paleobotany: An Introduction to Fossil Plant Biology. New York, New York: McGraw-Hill Book Company.

Thomas, Barry A., and Robert A. Spicer. 1987. The Evolution and Palaeobiology of Land Plants. London, England: Croom Helm. Portland, Oregon: Dioscoides Press.

Thomas, Roger D. K., and Everett C. Olson (editors). 1980. A Cold Look at the Warm-Blooded Dinosaurs. Boulder, Colorado: Selected Symposia Series, American Association for the Advancement of Science.

Tidwell, William D. 1975. Common Fossil Plants of Western North America. Provo, Utah: Brigham Young University Press.

Tiffney, Bruce H. (editor). 1985. Geological Factors and the Evolution of Plants. New Haven, Connecticut; London, England: Yale University Press.

Weishampel, David B., Peter Dodson, and Halszka Osmólska (editors). 1990. The Dinosauria. Berkeley, California: University of California Press.

Wilford, John Noble. 1987. The Riddle of the Dinosaur. New York, New York: Vantage Books.

Wright, H. E., Jr. (editor). 1983. Late-Quaternary Environments of the United States. Volume 2. The Holocene. Minneapolis, Minnesota: University of Minnesota Press.

Geology

Allègre, Claude. 1988. The Behavior of the Earth: Continental and Seafloor Mobility. Translated by Deborah Kurnes van Dan. Cambridge, Massachusetts: Harvard University Press.

Berglund, Bjorn E. (editor). 1986. Handbook of Holocene Palaeoecology and Palaeohydrology. New York, New York: John Wiley and Sons.

Frazier, William J., and David R. Schwimmer. 1987. Regional Stratigraphy of North America. New York, New York; London, England: Plenum.

Harris, Ann G., and Esther Tuttle. 1990. Geology of National Parks. Fourth Edition. Dubuque, Iowa: Kendall/Hunt Publishing Company.

Harris, David V., and Eugene P. Kiver. 1985. The Geologic Story of the National Parks and Monuments. Fourth Edition. New York, New York: John Wiley and Sons.

Imbrie, John, and Katherine Palmer Imbrie. 1979. Ice Ages: Solving the Mystery. Short Hills, New Jersey: Enslow Publishers. Reprinted 1986, Cambridge, Massachusetts: Harvard University Press.

McPhee, John. 1981. Basin and Range. New York, New York: Farrar, Straus and Giroux.

McPhee, John. 1986. Rising from the Plains. New York, New York: Farrar, Straus and Giroux.

Mahaney, William C. (editor). 1984. Quaternary Dating Methods. Developments in Palaeontology and Stratigraphy, 7. Amsterdam, Holland; New York, New York: Elsevier.

Nisbet, Euan G. 1987. The Young Earth: An Introduction to Archaean Geology. Boston, Massachusetts: Allen and Unwin.

Rutter, N. W. (editor). 1985. Dating Methods of Pleistocene Deposits and Their Problems. Geoscience Canada, Reprint Series 2.

Schopf, Thomas J. M. 1980. Paleoceanography. Cambridge, Massachusetts: Harvard University Press.

Stanley, Steven M. 1989. Earth and Life Through Time. Second Edition. New York, New York: W. H. Freeman and Company.

Biology

Buchsbaum, Ralph, Mildred Buchsbaum, John Pearce, and Vicki Pearce. 1987. Animals Without Backbones. Chicago, Illinois: The University of Chicago Press.

Eisenberg, John F. 1981. The Mammalian Radiations: An Analysis of Trends in Evolution, Adaptation, and Behavior. Chicago, Illinois: The University of Chicago Press.

Margulis, Lynn, and Karlene V. Schwartz. 1988. Five Kingdoms: An Illustrated Guide to the Phyla of Life on Earth. Second Edition. New York, New York: W. H. Freeman and Company.

Raven, Peter H., Ray F. Evert, and Helena Curtis. 1981. Biology of Plants. Third Edition. New York, New York: Worth Publishers, Inc.

Extinction

Larwood, G. P. (editor). 1988. Extinction and Survival in the Fossil Record. The Systematics Association Special Volume Number 34. Oxford, England: Clarendon Press.

Holland, H. D., and A. F. Trendall (editors). 1984. Patterns of Change in Earth Evolution. Berlin, West Germany; New York, New York: Springer-Verlag.

McGhee, George R., Jr. 1988. The Late Devonian extinction event: Evidence for abrupt ecosystem collapse. Paleobiology, volume 14, number 3, pp. 250–257.

Nitecki, Matthew H. (editor). 1984. Extinctions. Chicago, Illinois: The University of Chicago Press.

Raup, David M. 1986. The Nemesis Affair: A Story of the Death of Dinosaurs and the Ways of Science. New York, New York: W. W. Norton and Company.

Raup, David M. 1988. Extinction in the geologic past. Pp. 109– 119 in Donald E. Osterbrock and Peter H. Raven (editors), Origins and Extinctions. New Haven, Connecticut; London, England: Yale University Press.

Stanley, Steven M. 1984. Mass extinctions in the oceans. Scientific American, volume 250, number 6, pp. 64–72.

Stanley, Steven M. 1987. Extinction. New York, New York: Scientific American Library.

Human Paleontology

Bilsborough, A. 1990. Human Evolution. New York, New York: Chapman and Hall.

Johanson, Donald C., and Maitland A. Edey. 1981. Lucy: The Beginnings of Humankind. New York, New York: Simon and Schuster.

Johanson, Donald C., and James Shreeve. 1989. Lucy's Child: The Discovery of a Human Ancestor. New York, New York: William Morrow and Company Inc.

Gribbin, John R. 1982. The Monkey Puzzle: Reshaping the Evolutionary Tree. New York, New York: Pantheon Books.

Klein, Richard G. 1989. The Human Career: Human Biological and Cultural Origins. Chicago, Illinois: The University of Chicago Press.

Leakey, Richard E. 1981. The Making of Mankind. New York, New York: E. P. Dutton.

Lewin, Roger. 1987. Bones of Contention: Controversies in the Search for Human Origins. New York, New York: Simon and Schuster.

Lewin, Roger. 1988. In the Age of Mankind. A Smithsonian Book of Human Evolution. Washington, DC: Smithsonian Institution Press.

Smith, Fred H., and Frank Spencer (editors). 1984. The Origins of Modern Humans: A world Survey of the Fossil Evidence. New York, New York: A. R. Liss.

Tattersall, Ian, Eric Delson, and John A. Van Couvering (editors). 1988. Encyclopedia of Human Evolution and Prehistory. New York, New York: Garland Publishing.

Willis, Delta. 1989. The Hominid Gang: Behind the Scenes in the Search for Human Origins. New York, New York: Viking Press.

Journals, Series and Bibliographies

The most up-to-date information on and interpretation of fossils that is available is found in the periodic literature of science. The journals and series listed below are recommended because they contain a wide variety of articles and are readily available. Literature searches into specific topics of interest can be undertaken using bibliographies.

The Geology of North America—Decade of North American Geology Project
Canadian Journal of Earth Sciences
Journal of Geology

Journal of Paleontology
Journal of Vertebrate Paleontology
Paleobiology
Palaeontology
Quaternary Research
Treatise on Invertebrate Paleontology
Bibliography of Fossil Vertebrates (published by Society of Vertebrate Paleontology)
Bibliography and Index of Geology (published by the American Geological Institute)

Guide Books

The guides included below are, for the most part, geological in orientation, but fossils are included as an integral component within them. Hence the guides are a good reference for readers interested in paleontology.

Alt, David A., and Donald W. Hyndman. 1978. Roadside Geology of Oregon. Missoula, Montana: Mountain Press Publishing Company.

Alt, David A., and Donald W. Hyndman. 1984. Roadside Geology of Washington. Missoula, Montana: Mountain Press Publishing Company.

Alt, David A., and Donald W. Hyndman. 1986. Roadside Geology of Montana. Missoula, Montana: Mountain Press Publishing Company.

Arduino, Paolo, and Giorgio Teruzzi. 1986. Simon and Schuster's Guide to Fossils. New York, New York: Simon and Schuster.

Buchanan, Rex C., and James R. McCauley. 1987. Roadside Kansas: A Traveller's Guide to its Geology and Landmarks. Lawrence, Kansas: University of Kansas Press.

Casanova, Richard, and Ronald P. Ratkevich. 1981. An Illustrated Guide to Fossil Collecting. Third Edition. Happy Camp, California: Naturegraph Publishers.

Chronic, Halka. 1980. Roadside Geology of Colorado. Missoula, Montana: Mountain Press Publishing Company.

Chronic, Halka. 1983. Roadside Geology of Arizona. Missoula, Montana: Mountain Press Publishing Company.

Chronic, Halka. 1987. Roadside Geology of New Mexico. Missoula, Montana: Mountain Press Publishing Company.

Feldman, Robert. 1985. The Rockhound's Guide to Montana. Billings and Helena, Montana: Falcon Press Publishing Company, Inc.

Fritz, William J. 1985. Roadside Geology of the Yellowstone Country. Missoula, Montana: Mountain Press Publishing Company.

Gadd, Ben. 1986. Handbook of the Canadian Rockies: Geology, Plants, Animals, History and Recreation from Waterton/Glacier to the Yukon. Jasper, Alberta: Coax Press.

Lageson, David R., and Darwin R. Spearing. 1988. Roadside Geology of Wyoming. Missoula, Montana: Mountain Press Publishing Company.

Lindsay, Lowell, and Diana Lindsay. 1985. The Anza-Borrego Desert Region: A

Guide to the State Park and the Adjacent Areas. Second Edition. Berkeley, California: Wilderness Press.

Murray, Marian. 1967. Hunting For Fossils: A Guide to Finding and Collecting Fossils in all Fifty States. New York, New York: The Macmillan Company.

Orr, William N., and Elizabeth L. Orr. 1981. Handbook of Oregon Plant and Animal Fossils. Eugene, Oregon: Self-published.

Patton, Brian, and Bart Robinson. 1989. The Canadian Rockies Trail Guide—A Hiker's Manual to the National Parks. Revised Edition. Banff, Alberta: Summerthought Ltd.

Sheldon, R.A. 1979. Roadside Geology of Texas. Missoula, Montana: Mountain Press Publishing Company.

Stowe, Carlton H., and Lee I. Perry. 1979. Rockhound Guide to Mineral and Fossil Localities in Utah. Utah Geological and Mineral Survey, Utah Department of Natural Resources, Circular 63.

Thompson, Ida. 1982. The Audubon Society Field Guide to North American Fossils. New York, New York: Alfred A. Knopf.

Site Reports

The following is a list of the suggested readings that accompany the documentation of interpreted fossil sites. These publications are focused on specific fossils or specific geological time and place.

Agenbroad, Larry D., Jim I. Mead, and Lisa W. Nelson (editors). 1990. Megafauna and Man: Discovery of America's Heartland. Hot Springs, South Dakota: The Mammoth Site of Hot Springs, South Dakota, Inc. Scientific Papers, Volume 1.

Albritton, Claude C., Jr. 1980. The Abyss of Time: Changing Conceptions of the Earth's Antiquity after the Sixteenth Century. San Francisco, California: Freeman, Cooper and Company.

Allison, Ira S. 1966. Fossil Lake, Oregon: Its Geology and Fossil Faunas. Corvallis, Oregon: Oregon State University Press.

Averett, Walter R. (editor). 1987. Paleontology and Geology of the Dinosaur Triangle: Guidebook for the 1987 Field Trip, September 18–20, 1987. Grand Junction, Colorado: Museum of Western Colorado.

Bues, Stanley S., and Michael Morales (editors). 1990. Grand Canyon Geology. New York, New York; Oxford, England: Oxford University Press and Museum of Northern Arizona Press.

Bird, Roland T. 1985. Bones for Barnum Brown: Adventures of a Dinosaur Hunter. Fort Worth, Texas: Texas Christian University Press.

Bonneckson, Bill, and Rory M. Breckenridge (editors). 1982. Cenozoic Geology of Idaho. Idaho Bureau of Mines and Geology, Moscow, Bulletin 26.

Boyer, Bruce W. 1982. Green River laminates: Does the playa-lake model really invalidate the stratified-lake model? Geology, volume 10, pp. 321–324.

Breed, Carol S., and William J. Breed (editors). 1972. Investigations in the Triassic Chinle Formation. Museum of Northern Arizona Bulletin 47.

Callison, George, and Helen M. Quimby. 1984. Tiny dinosaurs: Are they fully grown? Journal of Vertebrate Paleontology, volume 3, number 4, pp. 200–209.

Camp, Charles L. 1980. Large ichthyosaurs from the Upper Triassic of Nevada. Palaeontographica, Abteilung A: Palaeozoologie-Stratigraphie, volume 170, number 4–6, pp. 139–200.

Colbert, Edwin H. 1968. Men and Dinosaurs. New York, New York: Dutton.

Colbert, Edwin H. 1984. The Great Dinosaur Hunters and Their Discoveries. New York, New York: Dover Publications Inc.

Colbert, Edwin H. 1989. The Triassic Dinosaur Coelophysis. Museum of Northern Arizona Bulletin 57.

Colbert, Edwin H., and R. Roy Johnson (editors). 1985. The Petrified Forest through the ages. Museum of Northern Arizona Bulletin 54.

Conway Morris, Simon (editor). 1982. Atlas of the Burgess Shale. London, England: Palaeontological Association.

Conway Morris, Simon, and Harry B. Whittington. 1985. Fossils of the Burgess Shale: A National Treasure in Yoho National Park, British Columbia. Geological Survey of Canada, Miscellaneous Report 43.

Currie, Philip J., Gregory C. Nadon, and Martin G. Lockley. 1991. Dinosaur footprints with skin impressions from the Cretaceous of Alberta and Colorado. Canadian Journal of Earth Sciences, volume 28, pp. 102–115.

Danis, Jane. 1988. Bibliography of vertebrate palaeontology in Dinosaur Provincial Park. Alberta: Studies in the Arts and Sciences, volume 1, number 1, pp. 225–234.

Darby, David G. 1982. The early vertebrate, Astraspis, habitat based on lithologic association. Journal of Paleontology, volume 56, number 5, pp. 1187–1196.

Dorf, Erling. 1964. The petrified forests of Yellowstone Park. Scientific American, volume 210, number 4, pp. 106–114.

Fagan, Brian M. 1987. The Great Journey: The Peopling of Ancient America. London, England; New York, New York: Thames and Hudson Ltd.

Farlow, James O. 1987. A Guide to Lower Cretaceous Dinosaur Footprints and Tracksites of Paluxy River Valley, Somerville County, Texas. Geological Society of America, South Central District, Baylor University.

Fischer, William A. 1978. The habitat of the early vertebrates: Trace and body fossil evidence from the Harding Formation (Middle Ordovician), Colorado. The Mountain Geologist, volume 15, number 1, pp. 1–26.

Foster, John, and Dick Harrison (editors). 1988. Tyrrell Museum of Palaeontology (a special commemorative issue featuring 20 separate articles). Alberta: Studies in the Arts and Sciences, volume 1, number 1.

Fritz, W. J. 1980. Reinterpretation of the depositional environment of the Yellowstone "fossil forests." Geology, volume 8, pp. 309–313.

Gillette, David D., and Martin G. Lockley (editors). 1989. Dinosaur Tracks and Traces. Cambridge, England; New York, New York: Cambridge University Press.

Gillette, David D., and David A. Thomas. 1985. Dinosaur tracks in the Dakota Formation (Aptian-Albian) at Clayton Lake State Park, Union County, New Mexico. Pp. 283–288 in Spencer G. Lucas and Jiri Zidek (editors). Santa Rosa Tucumcari Region, New Mexico Geological Society Guidebook, 36th Field Conference, Santa Rosa.

Gould, Stephen Jay. 1989. Wonderful Life: The Burgess Shale and the Nature of History. New York, New York; London, England: W. W. Norton and Company.

Grande, Lance. 1984. Paleontology of the Green River Formation, with a Review of the Fish Fauna. Second Edition. The Geological Survey of Wyoming, Bulletin 63.

Gross, Renie. 1985. Dinosaur Country: Unearthing the Badlands' Prehistoric Past. Saskatoon, Saskatchewan: Western Producer Prairie Books.

Harris, John M., and George T. Jefferson (editors). 1985. Rancho La Brea: Treasures of the Tar Pits. Natural History Museum of Los Angeles County, Science Series 31.

Hirsch, Karl F., and Betty Quinn. 1990. Eggs and eggshell fragments from the Upper Cretaceous Two Medicine Formation of Montana. Journal of Vertebrate Paleontology, volume 10, number 4, pp. 491–511.

Horner, John R. 1984. The nesting behavior of dinosaurs. Scientific American, volume 250, number 4, pp. 130–137.

Horner, John R., and James Gorman. 1988. Digging Dinosaurs. New York, New York: Workman Publishing.

Jackson, Richard W. 1980. The Fish of Fossil Lake: The Story of Fossil Butte National Monument. Jensen, Utah: Dinosaur Nature Association in cooperation with the National Park Service.

Jacobs, Louis L., and Phillip A. Murray. 1980. The vertebrate community of the Triassic Chinle Formation near St. Johns, Arizona. Pp. 55–71 in Louis L. Jacobs (editor), Aspects of Vertebrate History: Essays in Honor of E. H. Colbert. Flagstaff, Arizona: Museum of Northern Arizona Press.

Keener, James. 1989. Dinosaur Triangle: Land of the Terrible Lizards. Grand River, Colorado.

Kennedy, W. J., and W. A. Cobban. 1976. Aspects of Ammonite Biology, Biogeography, and Biostratigraphy. Special Papers in Palaeontology, Number 17. London, England: The Palaeontological Association.

Klein, Richard G., and Kathryn Cruz-Uribe. 1984. The Analysis of Animal Bones from Archeological Sites. Chicago, Illinois: The University of Chicago Press.

Kosch, Bradley F. 1990. A revision of the skeletal reconstruction of *Shonisaurus popularis* (Reptilia: Ichthyosauria). Journal of Vertebrate Paleontology, volume 10, number 4, pp. 512–514.

Koster, Emlyn H. 1987. Vertebrate taphonomy applied to the analysis of ancient fluvial systems. Pp. 159–168 in Frank G. Ethridge, Romeo M. Flores, and Michael D. Harvey (editors), Recent Developments in Fluvial Sedimentology, Society of Economic Paleontologists and Mineralogists, Special Publication Number 39.

Laury, Robert L. 1980. Paleoenvironment of a late Quaternary mammoth-bearing sinkhole deposit, Hot Springs, South Dakota. Geological Society of America Bulletin, Part I, volume 91, pp. 465–475.

Lehman, Ulrich. 1981. The Ammonites: Their Life and Their World. English Edition. Cambridge, England: Cambridge University Press.

Leopold, E. B., and H. D. MacGinitie. 1972. Development and affinities of Tertiary floras in the Rocky Mountains. Pp. 147–100 in Alan Graham (editor), Floristics and Paleofloristics of Asia and Eastern North America. Amsterdam, Holland; New York, New York: Elsevier Publishing Company.

Lockley, Martin G. 1986. The paleobiological and paleoenvironmental importance of dinosaur footprints. Palaios, volume 1, pp. 37–47.

Lockley, Martin G. 1986. Dinosaur Tracksites: A Guide to Dinosaur Tracksites of the Colorado Plateau and American Southwest. The First International Symposium on Dinosaur Tracks and Traces, Albuquerque, 1986. Geology Department Magazine, Special Issue Number 1, A University of Colorado at Denver, Geology Department Publication.

Lockley, Martin G. 1987. Dinosaur footprints from the Dakota Group of eastern Colorado. The Mountain Geologist, volume 24, number 4, pp. 107–122.

Long, Robert A., and Rose Houk. 1988. Dawn of the Dinosaurs: The Triassic in Petrified Forest. Petrified Forest, Arizona: Petrified Forest Museum Association.

Ludvigsen, Rolf. 1989. The Burgess Shale: Not in the shadow of the Cathedral Escarpment. Geoscience Canada, volume 16, number 2, pp. 51–59.

MacGinitie, H. D. 1953. Fossil Plants of the Florissant Beds, Colorado. Washington, DC: Carnegie Institution Publications, Number 599.

Madsen, J. H., Jr. 1976. *Allosaurus fragilis:* A Revised Osteology. Utah Geological and Mineralogical Survey, Bulletin 109.

Malley, Terry. 1987. Exploring Idaho Geology. Boise, Idaho: Mineral Land Publications.

Margulis, Lynn. 1988. The ancient microcosm of planet earth. Pp. 83–107 in Donald E. Osterbrock and Peter H. Raven (editors), Origins and Extinctions. New Haven, Connecticut; London, England: Yale University Press.

Martin, Larry D., and Debra K. Bennett. 1977. The burrows of the Miocene beaver *Palaeocastor*, western Nebraska, U.S.A. Palaeogeography, Palaeoclimatology, Palaeoecology, volume 22, pp. 173–193.

Martin, Paul S., and Richard G. Klein (editors). 1984. Quaternary Extinctions: A Prehistoric Revolution. Tucson, Arizona: The University of Arizona Press.

Massare, Judy A. 1988. Swimming capabilities of Mesozoic marine reptiles: Implications for method of predation. Paleobiology, volume 14, number 2, pp. 187–205.

Maxwell, Ross A. 1968. Big Bend of the Rio Grande: A Guide to the Rocks, Landscape, Geologic History, and Settlers of the Area of Big Bend National Park. Bureau of Economic Geology, The University of Texas at Austin, Guidebook F.

Maxwell, Ross A., John T. Lonsdale, Roy T. Hazzard, and John A. Wilson. 1967. Geology of Big Bend National Park, Brewster County, Texas. Austin, Texas: University of Texas, Bureau of Economic Geology Publication, Number 6711.

Newell, Norman D., J. Keith Rigby, Alfred G. Fischer, A. J. Whiteman, John E. Hickox, and John S. Bradley. 1953. The Permian Reef Complex of the Guadalupe Mountains Region, Texas and New Mexico. A Study in Paleoecology. San Francisco, California: W. H. Freeman and Company.

Pausé, P. H., and R. Gay Spears (editors). 1986. Geology of the Big Bend Area and Solitario Dome, Texas. West Texas Geological Society 1986 Field Trip Guidebook, Publication 86–82.

Raup, Omer B., Robert L. Earhart, James W. Whipple, and Paul E. Carrera. 1983. Geology along Going-to-the-Sun Road, Glacier National Park, Montana. Glacier Natural History Association.

Retallack, Greg J. 1983. Late Eocene and Oligocene Paleosols from Badlands National Park, South Dakota. Geological Society of America, Special Paper 193.

Russell, Dale A. 1989. An Odyssey in Time: The Dinosaurs of North America. Toronto, Ontario: University of Toronto Press.

Saenger, Walter. 1982. Florissant Fossil Beds National Monument—Window to the Past. Estes Park, Colorado: Rocky Mountain Nature Association, Inc.

Sternberg, Charles H. 1985. Hunting Dinosaurs in the Badlands of the Red Deer River, Alberta Canada. Third Edition, introduced by David A. E. Spalding. Edmonton, Alberta: NeWest Press.

Stock, Chester. 1956. Rancho La Brea: A Record of Pleistocene Life in California. Sixth Edition. Natural History Museum of Los Angeles County, Science Series 20, Paleontology Number 11.

Sutcliffe, Antony J. 1985. On the Track of Ice Age Mammals. Cambridge, Massachusetts: Harvard University Press.

Thayer, Dave. 1986. A Guide to Grand Canyon Geology along Bright Angel Trail. Grand Canyon, Arizona: Grand Canyon Natural History Association.

Ward, Peter Douglas. 1988. In Search of Nautilus: Three Centuries of Scientific Adventures in the Deep Pacific to Capture a Prehistoric-Living-Fossil. New York, New York: Simon and Schuster.

Whitaker, George O., and Joan Meyers. 1963. Dinosaur Hunt. New York, New York: Harcourt, Brace and World.

Whittington, Harry B. 1985. The Burgess Shale. New Haven, Connecticut: Yale University Press.

Wolfe, Jack. 1978. A paleobotanical interpretation of Tertiary climates in the northern hemisphere. American Scientist, volume 66, pp. 694–703.

Index

294

Taxonomic Index